WITHDRAWN

Compendium of Raspberry and Blackberry Diseases and Insects

Edited by

Michael A. Ellis
Ohio State University
Ohio Agricultural Research and Development Center, Wooster

Richard H. Converse
U.S. Department of Agriculture, Agricultural Research Service
Horticultural Crops Research Lab, Corvallis, Oregon

Roger N. Williams
Ohio State University
Ohio Agricultural Research and Development Center, Wooster

Brian Williamson
Scottish Crop Research Institute
Invergowrie, Dundee

APS PRESS
The American Phytopathological Society

Financial Sponsors

Ahrens Nursery and Plant Labs
CIBA-GEIGY Corporation
Mobay Corporation
Monsanto Company
Nourse Farms, Inc.

Cover photographs courtesy R. J. McNicol; copyright Scottish Crop Research Institute

Reference in this publication to a trademark, proprietary product, or company name by personnel of the U.S. Department of Agriculture or anyone else is intended for explicit description only and does not imply approval or recommendation to the exclusion of others that may be suitable.

Library of Congress Catalog Card Number: 91-76318
International Standard Book Number: 0-89054-121-3

© 1991 by The American Phytopathological Society

All rights reserved.
No portion of this book may be reproduced in any form, including photocopy, microfilm, information storage and retrieval system, computer database, or software, or by any means, including electronic or mechanical, without written permission from the publisher.

Copyright is not claimed in any portion of this work written by U.S. government employees as a part of their official duties.

Printed in the United States of America

The American Phytopathological Society
3340 Pilot Knob Road
St. Paul, Minnesota 55121-2097, USA

Preface

This compendium is designed as a practical reference for the identification, cause, epidemiology, and control of diseases and insect pests of raspberry and blackberry and their hybrids. Like other books in the series, it is intended to serve as a sourcebook for growers, crop advisors (from universities, agribusiness, and government), private consultants, students, and plant pathologists worldwide who work with the *Rubus* cane fruit industries. It is written for those with limited training in plant pathology, entomology, and horticulture, but it has sufficient detail to be of use to students and professionals alike. Each section was written by an active researcher or recognized expert on the subject.

Originally the compendium was intended to emphasize diseases of *Rubus* in North America; but its scope was expanded to include most of the major diseases worldwide. The section on insects was incorporated to make the book more useful to the production-oriented and pest control-oriented user.

This compendium was initiated from a need to update the U.S. Department of Agriculture's *Diseases of Raspberries and Erect and Trailing Blackberries*, (Agriculture Handbook No. 310), which was written by R. H. Converse and published in 1966. The handbook is the foundation upon which this compendium was developed.

Diseases caused by biotic agents, such as fungi, bacteria, viruses, and nematodes, are presented in Part I. Those caused by fungi are further subdivided according to the plant parts affected; viruses are subdivided according to method of transmission. Part II covers arthropod pests (insects and mites) of *Rubus*, and these are subdivided according to the plant parts they attack.

Because of the international scope of this compendium and the constantly changing regulations related to chemical control, our emphasis is on principles of disease and insect control rather than specific chemical control measures.

Disorders caused by abiotic factors, such as environmental stresses, nutrient deficiencies, and chemical toxicities, are discussed in Part III. Part IV examines how cultural practices can be used in disease management programs. Efforts to prevent disease spread in this clonally propagated crop through the development and use of disease-free planting material are described in Part V.

The authors who prepared the sections are identified at the ends of the sections. Sections without such acknowledgments were prepared by the editors. Acknowledgment of photographs and illustrations is given only where they were not provided by the authors of the accompanying text.

We are indebted to many individuals at the Ohio State University, Ohio Agricultural Research and Development Center, Wooster, including Jo Hershberger and Marilyn Snyder, for typing the manuscript; Cindy Gray, for artwork; and Ken Chamberlain and Margaret Latta, for photographic services.

We gratefully acknowledge the following individuals, who authored sections of the compendium. Those who reviewed portions of the manuscript as well are identified with an asterisk.

P. Ahrens, Ahrens Nursery, Huntingburg, Indiana
R. D. Akre, Department of Entomology, Washington State University, Pullman
D. Brayford, Commonwealth Agricultural Bureaux, International Mycological Institute, Kew, Surrey, England
P. R. Bristow,* Washington State University, Western Washington Research Center, Puyallup
R. H. Converse,* U.S. Department of Agriculture, Agricultural Research Service, Horticultural Crops Research Lab, Corvallis, Oregon
R. P. Davis,* Ministry of Agriculture, Fisheries, and Food, Harpenden, Hertfordshire, England
M. A. Ellis,* Department of Plant Pathology, Ohio State University, Ohio Agricultural Research and Development Center, Wooster
G. C. Fisher, Department of Entomology, Oregon State University, Corvallis
S. C. Gordon,* Scottish Crop Research Institute, Invergowrie, Dundee, Scotland
B. L. Goulart, Department of Horticulture, Pennsylvania State University, University Park
S. C. Gregory, Forestry Commission, Northern Research Station, Roslin, Midlothian, Scotland
W. D. Gubler,* Department of Plant Pathology, University of California, Davis
K. E. Hummer,* National Clonal Germplasm Repository, Corvallis, Oregon
D. T. Johnson,* Department of Entomology, University of Arkansas, Fayetteville
A. T. Jones,* Scottish Crop Research Institute, Invergowrie, Dundee, Scotland
W. C. Kleiner, Department of Plant Pathology, Pennsylvania State University, University Park
T. Locke, Ministry of Agriculture, ADAS, Wolverhampton, England
F. D. McElroy,* Peninsu-Lab, Kingston, Washington
J. L. Maas, U.S. Department of Agriculture Fruit Laboratory, Beltsville, Maryland
I. G. Montgomerie, Broughty Ferry, Dundee, Scotland
L. W. Moore,* Department of Botany and Plant Pathology, Oregon State University, Corvallis
N. L. Nickerson, Agriculture Canada Research Station, Kentville, Nova Scotia
H. S. Pepin,* Agriculture Canada, Vancouver, British Columbia
M. P. Pritts,* Department of Pomology, Cornell University, Ithaca, New York
S. M. Ries, Department of Plant Pathology, University of Illinois, Urbana
J. Rytter, Department of Plant Pathology, Pennsylvania State University, University Park
G. A. Schaefers, Department of Entomology, New York State Agricultural Experiment Station, Geneva
C. H. Shanks, Jr., Washington State University, Vancouver
B. J. Smith,* U.S. Department of Agriculture, Agricultural Research Service, Small Fruit Research Station, Poplarville, Mississippi
R. Stace-Smith,* Agriculture Canada Research Station,

Vancouver, British Columbia
B. C. Strik,* Department of Horticulture, Oregon State University, Corvallis
L. C. Stuart, Glengary, West Virginia
B. C. Sutton,* Commonwealth Agricultural Bureaux, International Mycological Institute, Kew, Surrey, England
J. W. Travis, Department of Plant Pathology, Pennsylvania State University, University Park
W. S. Washington, Plant Research Institute, Department of Agriculture and Rural Affairs, Burnley, Victoria, Australia
W. F. Wilcox,* Department of Plant Pathology, New York State Agricultural Experiment Station, Geneva
K. L. Wilder, Department of Soil Science, Oregon State University, Corvallis
R. N. Williams,* Department of Entomology, Ohio State University, Ohio Agricultural Research and Development Center, Wooster
B. Williamson,* Scottish Crop Research Institute, Invergowrie, Dundee, Scotland

We also wish to thank the following individuals, who reviewed sections of the manuscript:

P. F. Bertrand, University of Georgia, Tifton
D. J. F. Brown, Scottish Crop Research Institute, Invergowrie, Dundee, Scotland
T. J. Burr, Department of Plant Pathology, New York State Agricultural Experiment Station, Geneva
G. A. Cahoon, Department of Horticulture, Ohio State University, Ohio Agricultural Research and Development Center, Wooster
H. A. Daubeny, Agriculture Canada, Vancouver, British Columbia
J. Duncan, Scottish Crop Research Institute, Invergowrie, Dundee, Scotland
C. W. Ellett, Department of Plant Pathology, Ohio State University, Columbus
P. Fenn, Department of Plant Pathology, University of Arkansas, Fayetteville
M. Foley, ADAS, Cambridge, England
R. C. Funt, Department of Horticulture, Ohio State University, Columbus
G. J. Galletta, U.S. Department of Agriculture, Agricultural Research Service Fruit Laboratory, Beltsville, Maryland
M. O. Garraway, Department of Plant Pathology, Ohio State University, Columbus
E. G. Gray, Aberdeen, Scotland
G. Hall, Commonwealth Agricultural Bureaux, International Mycological Institute, Kew, Surrey, England
D. L. Jennings, Maidstone, Kent, England
R. K. Jones, Department of Plant Pathology, North Carolina State University, Raleigh
H. M. Lawson, Scottish Crop Research Institute, Invergowrie, Dundee, Scotland
J. F. Moore, Southwest Missouri State University, Fruit Experiment Station, Mountain Grove
J. N. Moore, Department of Horticulture, University of Arkansas, Fayetteville
J. E. M. Mordue, Commonwealth Agricultural Bureaux, International Mycological Institute, Kew, Surrey, England
M. C. M. Perombelon, Scottish Crop Research Institute, Invergowrie, Dundee, Scotland
J. Postman, National Clonal Germplasm Repository, Corvallis, Oregon
R. M. Riedel, Department of Plant Pathology, Ohio State University, Columbus
R. C. Rowe, Department of Plant Pathology, Ohio State University, Ohio Agricultural Research and Development Center, Wooster
E. Seemüller, Institut für Pflanzenschutz im Obstbau, Dossenheim, Germany
R. C. Shattock, University College of North Wales, Bangor
T. B. Sutton, Department of Plant Pathology, North Carolina State University, Raleigh
R. Watling, Royal Botanic Gardens, Edinburgh, Scotland

Additional photographs were supplied by the following individuals:

A. M. Agnello, Department of Entomology, New York State Agricultural Experiment Station, Geneva
P. F. Bertrand, University of Georgia, Tifton
J. M. Duncan, Scottish Crop Research Institute, Invergowrie, Dundee, Scotland
R. C. Funt, Department of Horticulture, Ohio State University, Columbus
S. Heady, Department of Entomology, Ohio State University, Ohio Agricultural Research and Development Center, Wooster
S. N. Jeffers, Department of Plant Pathology, University of Wisconsin, Madison
R. K. Jones, Department of Plant Pathology, North Carolina State University, Raleigh
T. L. Ladd, Jr., Department of Entomology, Ohio State University, Ohio Agricultural Research and Development Center, Wooster
D. Lorenz, Landes-Lehr und Forschungsanstalt für Landwirtschaft, Weinbau und Gartenbau, Neustadt an der Weinstrasse, Germany
L. V. Madden, Department of Plant Pathology, Ohio State University, Ohio Agricultural Research and Development Center, Wooster
R. J. McNicol, Scottish Crop Research Institute, Invergowrie, Dundee, Scotland
D. G. Nielsen, Department of Entomology, Ohio State University, Ohio Agricultural Research and Development Center, Wooster
M. C. Shurtleff, Department of Plant Pathology, University of Illinois, Urbana
A. Sivanesan, Commonwealth Agricultural Bureaux, International Mycological Institute, Kew, Surrey, England
R. Stace-Smith, Agriculture Canada, Vancouver, British Columbia
J. W. Travis, Department of Plant Pathology, Pennsylvania State University, University Park

Contents

Introduction
- 1 The Genus *Rubus*
- 2 The Raspberry and Blackberry Plants

Part I. Diseases Caused by Biotic Factors
- 3 **Cane and Foliar Diseases Caused by Fungi**
- 3 Anthracnose
- 5 Cane Blight
- 7 Midge Blight
- 7 Spur Blight
- 10 Cane Botrytis
- 11 Purple Blotch
- 12 Ascospora Dieback
- 12 Botryosphaeria Cane Canker of Blackberry
- 13 Rosette (Double Blossom)
- 15 Downy Mildew
- 16 Powdery Mildew
- 18 Raspberry Leaf Spot
- 18 Septoria Leaf Spot of Blackberry
- 19 Sydowiella and Gnomonia Cane Cankers
- 20 Nectria Canker of Raspberry
- 20 Silver Leaf
- 21 **Fruit and Flower Diseases Caused by Fungi**
- 21 Botrytis Fruit Rot (Gray Mold) and Blossom Blight
- 23 Postharvest Soft Rot (Leak Disease)
- 24 Minor Fruit Rots
- 25 Stamen Blight
- 26 **Rust Diseases**
- 26 Orange Rust
- 28 Cane and Leaf Rust
- 28 Yellow Rust
- 30 Late Leaf Rust
- 32 Blackberry Rust
- 33 Minor Rust Diseases
- 34 **Root and Crown Diseases Caused by Fungi**
- 34 Phytophthora Root Rot
- 36 Verticillium Wilt
- 37 Armillaria Root Rot
- 38 White Root Rot
- 39 **Diseases Caused by Bacteria**
- 39 Crown and Cane Gall
- 40 Fire Blight
- 41 Pseudomonas Blight
- 42 Leafy Gall
- 42 Hairy Root
- 42 **Diseases Caused by Viruses and Viruslike Agents**
- 43 **Aphid-Transmitted Diseases**
- 43 Raspberry Mosaic Disease Complex
- 45 Raspberry Leaf Curl
- 45 Cucumber Mosaic
- 46 Raspberry Vein Chlorosis
- 46 **Leafhopper-Transmitted Diseases**
- 46 Rubus Stunt
- 47 **Nematode-Transmitted Diseases**
- 47 European Nepovirus Diseases
- 49 Tomato Ringspot
- 51 **Pollen-Transmitted Diseases**
- 51 Raspberry Bushy Dwarf
- 52 **Viral Diseases with Unknown Methods of Natural Spread**
- 52 Apple Mosaic
- 53 Blackberry Calico
- 53 Cherry Leaf Roll
- 54 Tobacco Streak
- 55 Wineberry Latent Virus
- 56 Other Viruses and Viruslike Agents
- 59 **Nematode Parasites**
- 59 Root-Lesion Nematodes
- 60 Dagger and Needle Nematodes
- 62 Other Nematodes

Part II. Arthropod Pests
- 63 **Insects That Damage Roots and Crowns**
- 63 Raspberry Crown Borer
- 64 Root Weevils
- 64 Strawberry Crown Moth
- 65 **Insects That Damage Fruit**
- 65 **Scarab Beetles**
- 65 Japanese Beetle
- 65 Green June Beetle
- 66 Rose Chafer
- 66 Lygus Bugs
- 67 Raspberry Bud Moth
- 67 Picnic Beetles
- 68 Raspberry Fruitworms
- 69 Yellowjackets
- 69 Strawberry Bud Weevil (Clipper)
- 69 **Insects and Mites That Damage Foliage**
- 69 Spider Mites
- 70 Dryberry Mite (Raspberry Leaf and Bud Mite)
- 71 Redberry Mite
- 71 Raspberry Aphids
- 72 Leaf Rollers
- 73 Climbing Cutworms
- 73 Blackberry Psyllid
- 74 Western Winter Moth
- 74 Raspberry Sawflies and Leaf Miners
- 75 **Insects That Damage Canes**
- 75 Rednecked Cane Borer
- 75 Raspberry Cane Maggot
- 75 Raspberry Cane Midge
- 76 Raspberry Cane Borer

77 Tree Crickets
77 Rose Scale
78 Stalk Borer
78 **Insect Contaminants of Mechanically Harvested Fruit**

Part III. Disorders Caused by Abiotic Factors

79 **Nutritional Disorders**
80 **Herbicide Injury**
81 **Preemergent Herbicides**
82 **Postemergent Herbicides**
82 Translocatable Herbicides
82 Contact Herbicides
83 Volatile Herbicides
83 **Correcting Herbicide Excesses**
83 **Environmental Stress**
83 Temperature
84 Wind
84 Soil Moisture
85 Light
85 Solar Injury

Part IV. Effects of Cultural Practices on Disease

86 Using Disease- and Insect-Resistant Cultivars
86 Excluding and Reducing Pathogen Populations
87 Modifying Microclimate Within the Planting
88 Altering Production Practices to Prevent Plant Injury and Infection

Part V. Development of Healthy Planting Materials

89 Raspberry Certification Programs in North America
89 The Raspberry Certification Program in the United Kingdom
90 Nursery Production of Virus-Free Planting Material
90 International Exchange Regulations for *Rubus* Plant Material

93 **Glossary**

97 **Index**

Color Plates (following page **46**)

Introduction

The Genus *Rubus*

Raspberries and blackberries are a very diverse group of flowering plants in the genus *Rubus*. This genus is a member of the rose family (Rosaceae), which contains other genera prized for their horticultural qualities (apples, pears, cherries, plums, peaches, and strawberries). In North America, a common name for any member of the genus *Rubus* is *bramble*, which is derived from the Indo-European base word *bhrem*, meaning "coming to a point." This common name reflects the nearly universal trait of thorniness of the canes in most of these species. Blackberries generally have larger spines than raspberries, but spine density varies considerably, with some *Rubus* cultivars being completely thornless.

General characterizations of *Rubus* are difficult to make because of the diversity of plant growth habits and species distributions. Most species have perennial root systems and biennial canes; however, a few produce perennial canes, and others annual canes. Some produce fruit at the tops of first-year canes as well as on the lower portions of second-year canes. Most species are deciduous, but some are evergreen. Types of reproduction range from sexual to apomictic. Most species with edible fruit are found in mid-successional temperate climates (for example, at the edge of a forest or along a stream), but *Rubus* is present on every continent, on oceanic islands, and from tropical highlands to northern bogs above the Arctic Circle. Some species are vinelike, with prostrate canes and small leaves. Others are quite large, producing canes more than 11 m long. Their basic chromosome number is 7, but a range of ploidies exists. Most raspberries are diploid; blackberries range from diploid to dodecaploid.

The genus contains 12 subgenera, of which two—*R.* subg. *Idaeobatus* and *R.* subg. *Eubatus*—have attained major commercial significance as fruit crops. Several species of other subgenera are harvested in the wild for food, and occasionally species of these subgenera have been used in breeding programs. Two examples are cloudberry (*R. chamaemorus* L.) and arctic raspberry (*R. arcticus* L.), both important in alpine and circumpolar regions. Although their fruits resemble raspberries, these two species belong to *R.* subg. *Cylactis*, characterized by a dwarf, herbaceous, annual bearing habit. Hybrids of the European *R. arcticus* and the American *R. stellatus* Sm. are marketed in northern Europe.

Raspberries and blackberries produce aggregate fruit rather than true berries. The fruit develops from a single five-petaled flower, through the adhesion of many separate carpels (drupelets), which ripen simultaneously in one mass. The distinction between raspberries and blackberries is based on the adherence of the receptacle (torus) to the plant in ripened fruit: the receptacles stay attached to the raspberry plant but detach from the blackberry plant when the berries are harvested.

The worldwide distribution and production patterns of *Rubus* cane fruits will continue to change as new economic and trade policies are adopted and as technologies such as tissue culture propagation, virus indexing, gene manipulation, mechanical harvesting, herbicides, biological pest control, drip irrigation, and postharvest handling practices are introduced to help overcome production problems.

Raspberries

Raspberries are placed in *R.* subg. *Idaeobatus* to distinguish them from blackberries, which are in *R.* subg. *Eubatus* (Fig. 1). There are more than 200 recognized species of raspberries scattered throughout the world.

The major species that produce edible fruit include two ecotypes of *R. idaeus* L.—namely, *R. idaeus* subsp. *vulgatus* Arrhen. (the European red raspberry) and *R. idaeus* subsp. *strigosus* Michx. (the American red raspberry)—and *R. occidentalis* L. (the eastern North American black raspberry) and *R. glaucus* Benth. (a South American tetraploid black raspberry, which is likely a raspberry-blackberry hybrid).

Species that produce edible fruit but have no commercial significance are *R. leucodermis* Douglas ex Torr. & A. Gray (the western North American black raspberry); *R. spectabilis* Pursh (salmonberry); the Asiatic species *R. coreanus* Miq., *R. phoenicolasius* Maxim. (Japanese wineberry), *R. parvifolius* Nutt. (trailing raspberry), *R. ellipticus* Sm. (golden evergreen raspberry), *R. illecebrosus* Focke (strawberry raspberry), *R. kuntzeanus* Hemsley (Chinese raspberry), and *R. niveus* Thunb.; and the Hawaiian species *R. macraei* A. Gray and *R. hawaiiensis* A. Gray (akala berry).

Eastern Asia is considered the center of origin of *R.* subg. *Idaeobatus*.

Red raspberries are the most widely grown of all *Rubus* spp., although black raspberries are more popular in certain regions of the eastern United States. The progeny of crosses between black and red raspberries produce purple fruit and canes and are designated *R. neglectus* Peck. Purple raspberries exhibit hybrid vigor and resistance to insects and diseases, and they produce large fruit. The most successful hybrids are backcrosses of purple raspberry with red raspberry. Yellow-fruited *R. idaeus*, in which the fruit color is due to a recessive mutation, is also grown in small quantities for specialty markets. *R. occidentalis* genotypes with yellow fruit and with a primocane fruiting habit (bearing fruit on first-year canes) also exist, but neither is grown commercially.

Today, raspberries are grown throughout much of the world, with the Soviet Union accounting for a large percentage of total production. Other major production regions are in Europe (mostly in Poland, Hungary, Yugoslavia, Germany, and the United Kingdom) and the northern Pacific Coast (British Columbia, Washington, and Oregon). Many other countries (e.g., Chile, New Zealand, and Australia) have young industries with much production potential. World production in 1990 was estimated to be 375,000 t, on the assumption of a 5% annual increase in production. This rate of increase occurred during the 1980s and is expected to continue.

Blackberries

Blackberries are not as widely grown as raspberries, because of the limited adaptability of germ plasm with acceptable horticultural traits. Cultivated blackberries, for example, tend not to be as cold-hardy as raspberries, so their range is more restricted. Also, the wide availability of blackberries from the wild tended to delay the development of the commercial blackberry industry.

The abundance of wild blackberries exists because of certain traits of their life history. Most are colonists of habitats disturbed by either natural causes (fires, glaciers, etc.) or human activities (logging, farming, etc.). In fact, blackberries became a serious pest in Australia and South America after they had been introduced as boundary markers by early colonists. Sexual reproduction, dispersal of seed by birds, rapid vegetative propagation, and prolific production of apomictic seed have allowed extensive speciation in R. subg. *Eubatus*. Hybridization, polyploidy events, apomixis, and a highly variable phenotypic response to environment have made it difficult to classify blackberries in distinct biological species. Some taxonomists have defined only a few primary species with related hybrids, whereas others have estimated the number of species in the subgenus to be more than 100. Over the years, more than 5,000 Latin names of species have been used in attempts to classify blackberries. The variation in the subgenus might best be described as a network of relatively few sexual species and a great number of polyploid hybrid derivatives.

The most important domesticated blackberry in Europe is *R. laciniatus* Willd. It is now the most important cultivated species in the Pacific Northwest and also grows wild along the West Coast. The European species *R. nitidioides* Wats., *R. thyrsiger* Banning & Focke, and *R. rusticanus* E. Merc. have also been used in breeding programs. A cultivar of *R. procerus* P. J. Müll., native to northern Iran, was introduced by Luther Burbank under the name Himalaya Giant and has become a troublesome weed in Europe, North America, and New Zealand.

The primary species of blackberries in eastern North America seem to have intercrossed more than their European relatives, although higher ploidy types are more common in Europe. Sources of thornlessness, hardiness, disease resistance, flavor, fruit size, early ripening, heat tolerance, and productivity have been identified among blackberry species, including *R. allegheniensis* Porter (highbush blackberry), *R. canadensis* L. (mountain blackberry), *R. argutus* Link (tall blackberry), *R. cuneifolius* Pursh (sand blackberry), *R. frondosus* Bigel (leafy-flowered blackberry), *R. hispidus* L., *R. baileyanus* Britt., and *R. trivialis* Michx. (trailing dewberry). In western North America, *R. ursinus* Cham. & Schlechtend. and *R. macropetalus* Douglas are noted for their excellent flavor.

Blackberry-Raspberry Hybrids

Crosses between members of R. subg. *Idaeobatus* and R. subg. *Eubatus* succeed best at higher ploidy levels, because diploid hybrids are usually sterile. Blackberry-raspberry hybrids produce large amounts of high-quality fruit, but the fruit of some cultivars (such as tayberry) is difficult to pick, because the drupelets do not separate easily from the receptacle. Some of the more successful hybrids are tayberry, loganberry, boysenberry, and youngberry. Blackberries and blackberry-raspberry hybrids are intensively grown in the south central and northwestern United States, Europe, and New Zealand.

(Prepared by M. P. Pritts)

The Raspberry and Blackberry Plants

Plants of the genus *Rubus* typically have perennial roots and crowns and biennial shoots. The shoots (canes) generally have sharp spines, or thorns, and grow vegetatively during the first growing season and then become dormant during the winter. The following spring, they produce lateral branches, which flower and produce fruit. The entire cane dies after fruiting. Canes in their first growing season (in the vegetative state) are referred to as *primocanes*, and those in their second growing season (in the reproductive state) are *floricanes*. During the winter between their first and second growing seasons, the canes of most *Rubus* spp. require a "rest," or a dormant period. Raspberries require 800–1,600 hr of chilling at temperatures below 7°C (45°F) to satisfy their rest requirement. Blackberries generally have a shorter chilling requirement, of around 300–600 hr. At any given time, from budbreak in the spring until the removal of the dead floricanes in the late summer, primocanes and floricanes are both present in a *Rubus* planting. The extent of competition or support between the two types of canes has not been well established, but the presence of both provides the grower with challenges unique to *Rubus* production.

The various *Rubus* spp. differ markedly in many characteristics, including growth habit, the timing of fruiting, cold-hardiness, yield potential, fruit morphology and color, the presence or absence of spines, and disease susceptibility.

Selected References

Galletta, G. J., and Himelrick, D. G. 1990. Small Fruit Crop Management. Prentice-Hall, Englewood Cliffs, NJ. 602 pp.

Jennings, D. L. 1988. Raspberries and Blackberries: Their Breeding, Diseases and Growth. Academic Press, San Diego. 230 pp.

Pritts, M., and Handley, D. 1989. Bramble Production Guide. Northeast Regional Agricultural Engineering Service, Ithaca, NY. 189 pp.

(Prepared by B. L. Goulart)

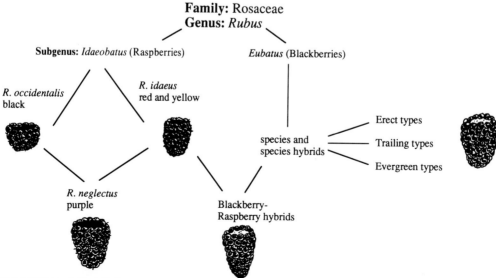

Fig. 1. Relationships between various *Rubus* spp.

Part I. Diseases Caused by Biotic Factors

Cane and Foliar Diseases Caused by Fungi

Anthracnose

Anthracnose was first reported in Italy in 1879 and in America in 1882. This disease, commonly called cane spot or gray bark, occurs in several species of *Rubus*. In North America, it is considered an extremely serious disease of black raspberry (*R. occidentalis* L.), red raspberry (*R. idaeus* L.), various raspberry hybrids, and occasionally blackberry (*R. fruticosus* L.). It is most important in black raspberry and susceptible cultivars of red raspberry. Severe yield losses may result from defoliation, wilting of lateral shoots, the death of fruiting canes, and damage from flower infection rendering fruit unmarketable.

Symptoms

Anthracnose symptoms are most conspicuous on canes but can also occur on leaves, petioles, pedicels, flower buds, and fruit. Reddish purple, circular to elliptical spots appear on primocanes in late spring. As the disease progresses, the spots enlarge and their centers become sunken, turning pale buff or ash gray, with the margins appearing slightly raised and purple (Plate 1). Most lesions appear in the middle cropping region of canes. Spots may merge, forming irregular blotches that encircle the cane. As lesions mature, they girdle the cane, causing it to dry out and crack (Plate 2). Cankered canes may die during winter, or they give rise to abnormal fruiting laterals with malformed fruit, especially in seasons of drought. Dieback of canes from the tip may also occur.

These early lesions on the cane are called pit lesions. Late summer and fall infections of the cane produce superficial lesions, in what is referred to as the gray bark phase. This phase is especially predominant on red raspberry, which overall has smaller and fewer cane lesions than black or purple raspberry. The gray appearance is due to the coalescence of many small surface colonies of the pathogen; these areas of coalesced colonies are approximately 3 mm in diameter. At this time, reddish, pimple-like acervuli are formed, usually arranged in concentric circles.

As on the canes, the first symptoms of infection on leaves are minute purple spots (1.5 mm in diameter), which later develop white centers (Plate 3). The diseased tissue may eventually drop out, causing a shot-hole effect. On blackberries and hybrid berries, the lesions may have wide purple margins, resembling a symptom of blackberry rust (caused by *Phragmidium violaceum* (C. F. Schulz) G. Wint.). Early-season infections of leaves of lateral shoots have totally destroyed the interveinal tissue and caused serious yield loss in Ireland.

Flowers and developing fruit may also be infected. Infected fruit has drupelets that remain small and are pitted and slow to ripen (Plate 4). The fruit is generally unfit for sale.

Causal Organism

Anthracnose is caused by the fungus *Elsinoe veneta* (Burkholder) Jenk. (anamorph *Sphaceloma necator* (Ellis & Everh.) Jenk. & Shear). The fungus is characterized by an extremely slow growth rate and sparse and variable production of conidia in culture. The hyphae are hyaline, septate, and branched and produce a bright red pigment. The fungus primarily overwinters as mycelium in infected canes. As it grows in the spring, it forms a stroma beneath the epidermal and subepidermal cells of the host. Eventually the stroma ruptures, and short, unbranched conidiophores arise from acervuli, bearing unicellular, hyaline conidia (5–7 × 2.5–3 μm) in a mucilaginous matrix. Conidia germinate and produce one or more germ tubes, which branch profusely, giving rise to additional stromatic structures. Subepidermal ascocarps (150 μm in diameter) are initiated in late summer on stromatic cushions; each contains several locules (Fig. 2), with one ascus per locule. The asci are globose (24–30 μm in diameter) and thick-walled. The ascospores (18–21 × 6–8 μm) are hyaline, four-celled, oblong elliptical, and borne eight per ascus in rows surrounded by a gelatinous sheath. Ascospores do not mature until the following spring. Ascocarps are produced only on canes.

Disease Cycle and Epidemiology

In the spring, conidia and ascospores are liberated from fruiting structures produced on canes infected during the previous season (Fig. 3). Observations in Wisconsin early this century indicated that ascospores can serve as primary inoculum; however, conidia are generally considered the most important source of primary inoculum in the spring. Ascospores are forcibly discharged during damp periods in late

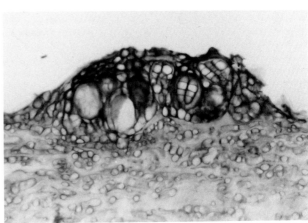

Fig. 2. Cross section of pseudothecium of *Elsinoe veneta*, containing four-celled ascospores. (Courtesy B. Williamson; copyright Scottish Crop Research Institute. Used by permission)

spring, and conidia are splash-dispersed during rain. Both types of spores germinate in 3–12 hr and infect only very young green tissues.

Inoculation experiments in Scotland showed that anthracnose lesions may take 4–6 weeks to develop. Early in the development of the disease, the primary cortex of the cane is extensively colonized and eventually collapses. Hyperplasia of parenchyma cells occurs in the infected area, which at first causes the lesions to appear raised. This tissue eventually collapses, and the area becomes sunken. It is in this region that the stroma forms. The pathogen continues its invasion, penetrating the main vascular system between the vascular bundles and phloem tissue. Xylem is not usually affected by the fungus. The stroma matures and erupts, and an acervulus develops in its place. Conidia and ascospores are produced and released to begin another infection cycle.

The primary damage to plants is caused by early infections. If the fungus infects the cane when the vascular tissues are immature, normal development of the phloem and xylem stops. Mature phloem fibers are resistant to the pathogen. Infections that occur later in the growing season are not as serious and usually cause smaller, shallower lesions.

Control

In areas where anthracnose is a problem, overhead irrigation should be avoided, to prevent dispersal of the pathogen. All steps possible should be taken to improve air circulation within a planting, to allow faster drying of foliage and canes. Reducing the number and duration of wet periods should reduce the potential for infection. Excessive applications of fertilizer (especially nitrogen) should be avoided, since it promotes excessive growth of very susceptible succulent plant tissue. Plants should be maintained in narrow rows (40–60 cm wide) and thinned to improve air circulation and allow better light penetration. Weeds are very effective in reducing air movement; therefore, good weed control within and between rows, as well as the removal of suckers, is important for improving air circulation within a planting.

Historically, pathologists have recommended the removal of cane stubs, or handles, from dormant plants to prevent the introduction of the pathogen into new plantings. This form of pathogen introduction should not be a problem if growers use plants produced by meristem tip culture, in which they are propagated and increased in the absence of the disease.

Because the fungus overwinters on both living and dead plant tissues, old fruiting canes and infected primocanes should be removed from the planting after harvest and destroyed (burned or buried). This greatly reduces the amount of overwintering inoculum that survives within the planting.

Wild brambles serve as an excellent source of inoculum for anthracnose and most other diseases. The area surrounding the planting should be kept free of wild blackberries and raspberries.

If anthracnose is established within a planting and regularly causes some damage, fungicides may be required for satisfactory control. Early-season fungicide sprays are often necessary. Liquid lime sulfur applied at the end of the dormant period may eradicate a portion of the overwintering inoculum. The timing of the application is critical: the fungicide should be applied during the period from the green tip stage of plant development (when the plant is breaking dormancy) to the point at which no more than 15 mm of green tissue has been formed. Sprays prior to green tip are ineffective against the pathogen, and later sprays may cause phytotoxicity in young foliage. Additional fungicide applications have generally been recommended when primocanes are 15–20 cm tall and again at 14-day intervals through harvest. Since the fungus requires free water for infection, fungicides can be withheld during extended periods of dry weather.

Commonly used fungicides for anthracnose control include captan, benomyl, dichlofluanid, and ferbam. Fungicides applied during bloom to control gray mold (caused by *Botrytis*

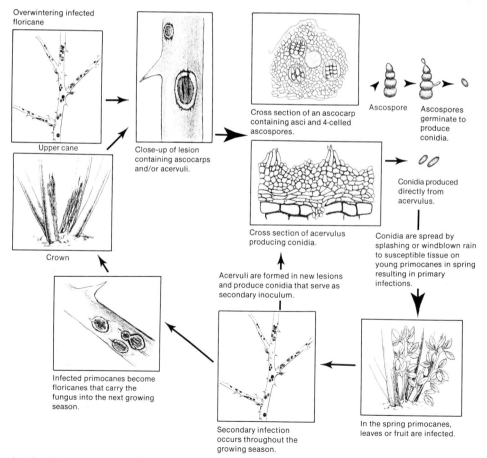

Fig. 3. Disease cycle of anthracnose, caused by *Elsinoe veneta*. The fungus overwinters in bark or within lesions on old fruited canes and new floricanes as ascocarps or acervuli, which produce ascospores and conidia, respectively, in the spring.

cinerea Pers.:Fr.) are generally effective against *E. veneta*. However, strains of *E. veneta* with resistance to MBC-generating fungicides (benomyl, thiophanate-methyl, and carbendazim) have been detected in Scotland, and the fungus has been shown to be insensitive to the dicarboximide fungicides (vinclozolin and iprodione). When fungicides are used, it is important to obtain complete coverage of the entire plant.

Strong host resistance has been found in several red raspberry cultivars, including Willamette, Heritage, Chilcotin, Nootka, Meeker, and Autumn Bliss.

Selected References

Burkholder, W. H. 1917. The anthracnose of the raspberry and related plants. Pages 155-183 in: N.Y. (Cornell Univ.) Agric. Exp. Stn. Bull. 395.

Harris, R. V. 1931. Raspberry cane spot: Its diagnosis and control. J. Pomol. Hortic. Sci. 9:73-99.

Harris, R. V. 1933. The infection of raspberry fruits by the cane spot fungus. Pages 86-89 in: Rep. East Malling Res. Stn. 1932.

Labruyère, R. E. 1957. Observations on *Elsinoe veneta* (Burkh.), the perfect-form of *Sphaceloma* (Ell. & Ev.) on raspberry. Tijdschr. Plantenziekten 63:153-158.

Munro, J. M., Dolan, A., and Williamson, B. 1988. Cane spot (*Elsinoe veneta*) in red raspberry: Infection periods and fungicidal control. Plant Pathol. 37:390-396.

Ó Ríordáin, F. 1969. Control of cane spot and spur blight of raspberry with fungicides. Pages 155-158 in: Proc. Br. Insectic. Fungic. Conf. 5th.

Sivanesan, A., and Critchet, C. 1969. *Elsinoe veneta*. Descriptions of Pathogenic Fungi and Bacteria, No. 484. Commonwealth Mycological Institute, Kew, Surrey, England.

Williamson, B., Hof, L., and McNicol, R. J. 1989. A method for *in vitro* production of conidia of *Elsinoe veneta* and the inoculation of raspberry cultivars. Ann. Appl. Biol. 114:23-33.

Williamson, B., and McNicol, R. J. 1989. The histology of lesion development in raspberry canes infected by *Elsinoe veneta*. Ann. Appl. Biol. 114:35-44.

(Prepared by J. W. Travis and J. Rytter)

Cane Blight

Cane blight was first described in New York in 1902. The disease only affects canes that have been wounded in their vegetative year. It sporadically attacks canes of all *Rubus* spp., causing vascular damage during the winter and then bud failure, lateral shoot wilt, or cane death the following spring. Recent information about cane blight was gained through research on raspberries harvested by machine in Scotland, in which the disease has been particularly severe.

Symptoms

All symptoms of cane blight occur in close association with wounds. External symptoms are usually not visible on primocanes. If the epidermis of an infected primocane is scraped off to expose the vascular tissues during late fall, a brown stripe lesion can be detected spreading from a wound (Plate 5). By spring, a lesion may extend through several internodes on one side of the cane, causing the death of axillary buds or wilt of individual lateral shoots as it continues to spread. Lesions can girdle the vascular tissues during winter, causing cane death. If cane girdling occurs in late spring or summer, the entire cane above the infected wound may wilt and die suddenly. The pathogen removes cellulose from phloem fibers, phloem, and xylem near the wound. This causes the cane to become brittle at the point of infection, so that it can easily be broken by slight pressure. The epidermis often develops a silver color by early spring, and dark gray patches of dried conidia are exuded around immersed pycnidia (Plate 6). These symptoms are common on stubs of old fruiting canes left around the crown after pruning.

On blackberry stems, which usually remain green throughout the winter, the lesions are visible as dark red to purple areas around wounds, with irregular dark purple borders. In time, the lesion centers usually become grayish. These lesions are similar to those of Botryosphaeria cane canker.

Causal Organism

Cane blight is caused by the fungus *Leptosphaeria coniothyrium* (Fuckel) Sacc. (anamorph *Coniothyrium fuckelii* Sacc.). This pathogen also causes stem canker of roses and diseases of several other woody perennials. The immersed, subglobose pseudothecia (250-350 μm in diameter) are found only on the dead stubs of old floricanes (Fig. 4). They develop bitunicate asci (60-70 × 6-7 μm) containing eight pale olive brown, three-septate ascospores (12-15 × 3.5-4.5 μm).

The immersed, subglobose pycnidia (200-260 μm in diameter) produce globose, single-celled conidia (2.5-4 μm), which in mass appear olivaceous to pale brown by transmitted light. When discharged and dried on canes, conidial masses are silvery gray.

Disease Cycle

Ascospores mature in late April and May; there is little information about their infectivity. Conidia are exuded from pycnidia and dispersed by splashing rain from early spring to late fall (Fig. 5). Inoculation experiments suggest that the age of tissues and the time of wounding are important for infection and disease development. The canes of red raspberries became highly resistant to infection in the fall. The pathogen occurs on numerous woody hosts, but evidence indicates that old infected floricanes are the primary source of inoculum in *Rubus* plantings.

The fungus requires a wound or tissue damage in order to enter the vascular tissue of the plant. Unwounded canes can be infected after inoculation with mycelium, but penetration of the periderm takes several months, and vascular damage is generally slight. It is therefore unlikely that the disease would cause significant damage or yield loss without

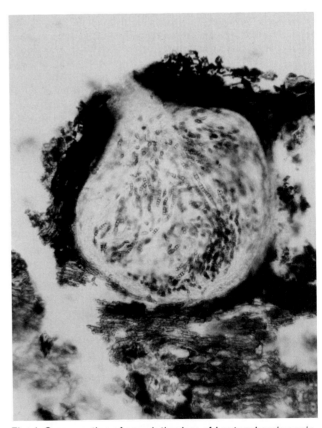

Fig. 4. Cross section of pseudothecium of *Leptosphaeria coniothyrium*, showing dark-walled, four-celled ascospores. (Courtesy B. Williamson; copyright Scottish Crop Research Institute. Used by permission)

wounds. Wounds made by abrasion against old cane stubs and spines are common infection courts, and those made by hoeing, farm machinery, and mechanical harvesters and sites of serious herbicide damage are also common points of entry. Wounds made when primocanes are still young (before July) become infected rapidly, and the canes may collapse within a month. These early losses are generally unimportant unless a high proportion of canes are affected.

The infection of wounds made during harvest is the most economically significant in Scotland. Infections that occur at this time often remain symptomless until after fall. Growers, assuming the canes are healthy, prune plantings to the correct cane density. In spring, the damage caused by the disease becomes evident when lateral shoots fail to grow. It is then too late to compensate for cane loss.

Control

Cane blight has caused serious yield losses in growing seasons following heavy rainfall before or during harvest, or following intensive overhead irrigation. Any practices that reduce the risk of splash dispersal of conidia are beneficial in control. Overhead irrigation should be minimized during critical times for infection. As in the control of most other *Rubus* diseases caused by fungi, it is very important to create an open plant habit that allows greater air circulation and sunlight penetration to hasten the drying of the plants after rain or irrigation.

A major consideration in disease control is the prevention of damage to or wounding of primocanes. This is generally achieved by adjustments in all practices involving the movement of machinery and people through the planting.

Studies of several mechanical harvesters in Scotland have shown that picking heads with plastic beater bars or rubber-covered vibrating fingers result in much less wounding of canes than harvesters with overlapping spring-loaded catching plates (fishplates), which cause serious cane damage. In addition, it is likely that the catchers introduce inoculum into new wounds they make. Thus, significant disease control can be obtained by making some relatively minor adjustments in mechanical harvesting machinery. In Scotland, improved design of catching devices has reduced the incidence of cane blight. The disease apparently causes little damage in mechanically harvested fields in the Pacific Northwest, probably due to drier weather during harvest.

Cane vigor control (cane burning) has reduced the damage to primocanes caused by harvesting operations and reduced the severity of cane blight. Biennial cropping (alternate-year bearing), in which no primocanes are present in the year in which fruit is picked, helps to avoid the disease. Other cane-training methods that segregate primocanes from floricanes during picking can reduce cane blight severity. However, care should be taken not to create a dense canopy, which is difficult to spray efficiently and may increase drying time in the row.

High-volume sprays of benomyl or thiophanate-methyl applied before, during, or immediately after harvest have given good control of cane blight in Scotland, and no resistant strains of *L. coniothyrium* have been detected. It is important that the bottom half of all canes receive adequate fungicide coverage. In several locations in North America, the application of a fungicide (generally benomyl) following any pruning during the growing season, such as topping black raspberries, has been recommended. If fungicide is applied in this manner, it should be done after pruning but before rain or irrigation.

The American cultivar Latham shows some tissue resistance in wound inoculation tests, and strong resistance has been identified in *R. pileatus* Focke and in its hybrids with red raspberry.

Selected References

Jennings, D. L. 1979. Resistance to *Leptosphaeria coniothyrium* in red raspberry and some related species. Ann. Appl. Biol. 93:319-326.

Punithalingam, E. 1980. *Leptosphaeria coniothyrium*. Descriptions of Pathogenic Fungi and Bacteria, No. 663. Commonwealth Mycological Institute, Kew, Surrey, England.

Ramsay, A. M., Cormack, M. R., Mason, D. T., and Williamson, B. 1985. Problems of harvesting raspberries by machine in Scotland—A review of progress. Agric. Engineer 40:2-9.

Seemüller, E., Kartte, S., and Erdel, M. 1988. Penetration of the periderm of red raspberry canes by *Leptosphaeria coniothyrium*. J. Phytopathol. 123:362-369.

Stewart, F. C., and Eustace, H. J. 1902. Raspberry cane blight and raspberry yellows. I. Raspberry cane blight. Pages 333-362 in: N.Y. Agric. Exp. Stn. Bull. 226.

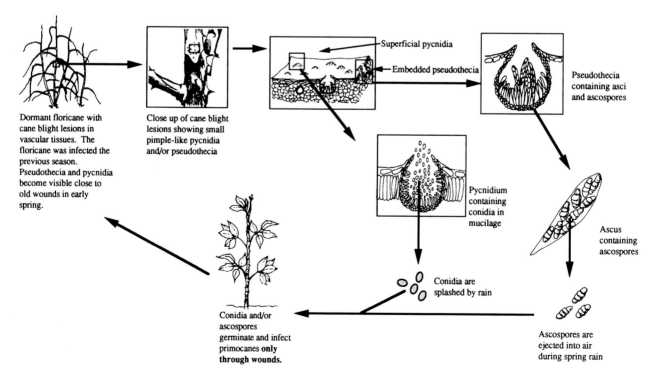

Fig. 5. Disease cycle of cane blight, caused by *Leptosphaeria coniothyrium*. The fungus overwinters on dead floricanes and on floricanes infected the previous season as pseudothecia that produce ascospores or pycnidia that produce conidia. These ascospores and conidia serve as primary inoculum in the spring. (Courtesy M. A. Ellis; drawing by Cindy Gray)

Williamson, B., Bristow, P. R., and Seemüller, E. 1986. Factors affecting the development of cane blight (*Leptosphaeria coniothyrium*) on red raspberries in Washington, Scotland and Germany. Ann. Appl. Biol. 108:33-42.

Williamson, B., and Hargreaves, A. J. 1978. Cane blight (*Leptosphaeria coniothyrium*) in mechanically harvested red raspberry (*Rubus idaeus* L.). Ann. Appl. Biol. 88:37-43.

Williamson, B., and Hargreaves, A. J. 1981. The effect of sprays of thiophanate-methyl on cane diseases and yield in red raspberry, with particular reference to cane blight (*Leptosphaeria coniothyrium*). Ann. Appl. Biol. 97:165-174.

Williamson, B., and Ramsay, A. M. 1981. Prospects for control of cane blight in machine-harvested raspberries. Pages 281-285 in: Proc. Crop Prot. North. Britain 1981.

Williamson, B., and Ramsay, A. M. 1984. Effects of straddle-harvester design on cane blight (*Leptosphaeria coniothyrium*) of red raspberry. Ann. Appl. Biol. 105:177-184.

(Prepared by B. Williamson)

Midge Blight

Midge blight is a disease complex involving several fungal pathogens and the larvae of the raspberry cane midge (*Resseliella theobaldi* (Barnes)). It is described here as a cane disease because it is very similar to cane blight. (For details of the life cycle of *R. theobaldi*, see Arthropod Pests.)

Midge blight is confined to Europe and attacks mainly red raspberry cultivars, which produce natural growth splits in the surface of primocanes. The production of these splits in relation to the emergence of successive generations of adult midges is the predominant predisposing factor in this disease complex. Cane losses of 50-90% have been recorded in areas where the onset of splitting coincides precisely with midge emergence. Damage generally results from oviposition (egg laying) by the midge in growth splits.

Symptoms

Primocanes develop small, narrow growth splits, often beneath the swollen leaf bases at nodes up to 30 cm above the soil level in early spring. First-generation midge larvae damage tissues inside these splits, and fungi invade the damaged tissue. Affected splits may become dark brown or purple, and by midsummer they develop into deeply furrowed brown cankers. On either side of the cankers, ridges form as a result of increased cambial activity. In the center of each canker, the vascular tissues are destroyed, and the phloem fibers are visible. These cankers arise from rapid fungal invasion of canes damaged by the larvae. The cankered canes are weakened and may break in high winds or during harvest. Yield losses from cane breakage are rarely important, because a minority of the primocanes are affected. There have been no detailed studies of the fungal species responsible for this first-generation damage, but it is assumed that the infections are due to the same species causing damage following infestation by second-generation larvae.

Later in the season, second-generation midge larvae are present under the cane surface at natural splits. Several weeks after the larvae mature and drop to the soil, dark brown, lobate patch lesions become visible on the dark green surface of the newly exposed cork (periderm) in the splits (Plate 118). Superficially, the lesions may become blackened and have a rough appearance as fungal pycnidia develop; they usually turn silver later in the winter. In some cases, they may appear salmon pink on the surface, from the sporulation of *Fusarium* spp. Patch lesions penetrate deeply into vascular tissues and can be easily observed by scraping the canes to remove the epidermis and cork during winter or the following spring. The irregular, sharply defined patch lesions have been shown to be caused by fungal invasion of feeding sites of second-generation midge larvae and are a reliable indication that the pest is present in the planting. Some scraped canes with patch lesions may also show spreading brown stripe lesions. The incidence and severity of stripe lesions varies with the site and the year.

In spring, floricanes that were severely infested by the cane midge in the previous growing season may suffer some bud failure, wilt of lateral shoots, or complete death. The extent of symptom development depends on the area and extent of vascular tissue destroyed by fungal infection. Following a severe attack of midge blight, as in cane blight, infected primocanes may continue to grow normally and appear healthy.

Causal Organisms

The feeding damage caused by the second-generation larvae of *R. theobaldi* predisposes canes to colonization by a range of fungi. *Didymella applanata* (Niessl) Sacc. or *F. avenaceum* (Fr.:Fr.) Sacc. may sporulate on the surface and have been isolated from the vascular tissues beneath larval feeding sites, along with a range of other fungi, including *Phoma macrostoma* Mont. var. *macrostoma*, *F. culmorum* (Wm. G. Sm.) Sacc., and *Alternaria* spp. These fungi apparently do not spread beyond the edge of the feeding site. However, when *Leptosphaeria coniothyrium* (Fuckel) Sacc. invades feeding sites, spreading brown vascular lesions extending both up and down from the site of infection generally occur. Pycnidia of the anamorph *Coniothyrium fuckelii* Sacc. may develop on the surface of the affected area. Isolation of *L. coniothyrium* from these lesions is generally more successful in tissues close to the point of infection than in tissues at the margins of spreading lesions.

Control

Although fungi are responsible for the vascular damage, fungicides have invariably failed to control midge blight. In some cases, MBC-generating fungicides, such as benomyl, have increased the incidence and severity of the disease, perhaps by affecting the growth rate of canes and their tendency to produce natural splits.

Good control is achieved by controlling the raspberry cane midge. Insecticides timed to kill first-generation midges are the most effective means of control. Cane vigor control (cane burning) is also beneficial.

Selected References

Gordon, S. C., and Williamson, B. 1984. Raspberry cane blight and midge blight. Minist. Agric. Fish. Food (G.B.) Leafl. 905.

Pitcher, R. S., and Webb, P. C. R. 1952. Observations on the raspberry cane midge (*Thomasiniana theobaldi* Barnes). II. 'Midge blight,' a fungal invasion of raspberry cane following injury by *T. theobaldi*. J. Hortic. Sci. 27:95-100.

Williamson, B. 1984. Problems of diagnosis and control of raspberry cane blight and midge blight in Scotland. Pages 264-369 in: Proc. Crop Prot. North. Britain.

Williamson, B. 1987. Effect of fenitrothion and benomyl sprays on raspberry cane midge (*Resseliella theobaldi*) and midge blight, with particular reference to *Leptosphaeria coniothyrium* in the disease complex. J. Hortic. Sci. 72:171-175.

Williamson, B., and Hargreaves, A. J. 1979. A technique for scoring midge blight of red raspberry, a disease complex caused by *Resseliella theobaldi* and associated fungi. Ann. Appl. Biol. 91:297-301.

Williamson, B., and Hargreaves, A. J. 1981. Fungi on red raspberry from lesions associated with feeding wounds of raspberry cane midge (*Resseliella theobaldi*). Ann. Appl. Biol. 91:303-307.

Williamson, B., Lawson, H. M., Woodford, J. A. T., Hargreaves, A. J., Wiseman, J. S., and Gordon, S. C. 1979. Vigor control, an integrated approach to cane, pest and disease management in red raspberry (*Rubus idaeus*). Ann. Appl. Biol. 92:359-368.

(Prepared by S. C. Gordon and B. Williamson)

Spur Blight

Spur blight is a disease specific to *Rubus*. It affects black raspberry, blackberry, and blackberry-red raspberry hybrids worldwide. Red raspberry is most seriously affected, particularly in the Pacific Northwest and in Europe. In extremely vigorous or overgrown and weed-infested plantings the disease causes yield losses, especially if excessive nitrogen has been applied.

The axillary buds at infected nodes are substantially smaller and less capable of growing into lateral shoots (spurs) during the fruiting years than those at uninfected nodes. In Scotland, these buds remain alive but physiologically suppressed by healthy buds higher on the cane. This bud failure is the principal cause of yield loss if a high proportion of nodes in the upper cropping region of canes is affected. Reports from Poland indicate that affected canes may be more vulnerable to winter injury than uninfected ones.

Symptoms

Infections on leaves of primocanes are initiated at the leaf margin and advance inward toward the midvein. This results in a brown, V-shaped lesion with broad yellow margins (Plate 7). The fungus can attack foliage on lateral shoots of floricanes, but this is rare. The infection eventually spreads from the leaf through the petiole and into the node. Infected leaves are usually shed prematurely. A dark chestnut brown, spreading lesion develops on the cane below the node and around the axillary buds, and lesions from adjacent nodes often merge in the internode. Spur blight lesions often appear purple on some cultivars that have a heavy wax bloom; if the wax is removed by gentle rubbing, the lesions become dark brown (Plate 8). Lesions of cane Botrytis, caused by *Botrytis cinerea* Pers.:Fr., affect primocanes in a similar manner but can be distinguished from spur blight by their pale brown color. Spur blight lesions become indistinct and nearly undetectable later in the fall, as primocanes turn brown after cork layers mature in preparation for winter. During the winter, silver or gray lesions appear (Plate 9), and small black pseudothecia, and later pycnidia, develop on them; in some cultivars these lesions may be poorly developed and difficult to see.

The disease affects only the primary cortex of the cane, which lies outside the deep-seated cork layers. If the lesions are scraped, healthy green vascular tissues are exposed beneath the cork.

Buds at infected nodes may fail to grow in the spring, or they may develop into apparently healthy lateral shoots. A reduction in the number of flowers per infected node on apparently healthy lateral shoots from infected nodes has been recorded in British Columbia, but not in Scotland.

Causal Organism

Spur blight is caused by *Didymella applanata* (Niessl) Sacc. (the anamorph is an unnamed *Phoma* sp.). The immersed, subglobose pseudothecia (200–270 μm in diameter) have erumpent black ostioles and develop bitunicate, cylindrical to subclavate, eight-spored asci (60–75 × 10–15 μm) arranged in a relatively flat layer. The two-celled, hyaline ascospores (12–18 × 5–7 μm) are slightly ellipsoidal; the upper cell is broader than the lower (Fig. 6). The subglobose pycnida (200–250 μm) develop in clusters among the pseudothecia and are superficially indistinguishable from them. The pycnidia produce hyaline, cylindrical, one-celled conidia (4–7 × 2–3.5 μm).

Disease Cycle and Epidemiology

The fungus survives as mycelium, pseudothecia, and pycnidia on infected canes during winter (Fig. 7). Under wet conditions during summer, it produces ascospores and conidia, which infect the leaves of primocanes. Lesions first appear in June or July at the lowermost nodes of primocanes. Nodes become infected progressively higher on canes, but usually only the lower third or half of the primocane is attacked. Observations in Scotland suggest that infection of the leaf petiole may retard the development of the adjacent axillary bud. This effect is more pronounced in early-season than late-season infections.

Ascospores are discharged from April to August, usually with a maximum discharge in May. Mature or slightly senescent leaves are susceptible, but young leaves appear to be resistant. Conidia that are splash-dispersed during heavy rain in July and August have been correlated with the appearance of nodal lesions in England. These conidia are probably the most important inoculum of the fungus. As light is excluded from the base of canes later in the growing season, particularly in vigorous plantations, the oldest (lower) leaves become susceptible at a time when conidia are abundant and conditions for infection favorable.

D. applanata produces longer lesions and more abundant pseudothecia and pycnidia on infected canes during warm springs than on those grown under colder conditions. Prolonged mild temperatures in the fall may increase the likelihood of bud failure at infected nodes the following spring.

In Ontario, *D. applanata* can attack the bud scales, causing scale separation and desiccation and death of buds in winter. In Scotland, buds at infected nodes generally remain green and viable. Histological studies of naturally infected nodes in Scotland have shown that the infection route to the buds from the petiole is usually blocked by a primary protective layer of suberized and lignified cells across the adaxial cortex of the petiole. Such a layer is absent on the abaxial side, and hence the rapid colonization of the cane cortex below the leaf.

Two pathotypes that differ in their ability to infect the cultivar Newburgh have been described in eastern Canada. In similar studies on several cultivars in England, no pathogenic variation indicative of pathotypes was observed.

In Europe, pycnidia of *D. applanata* are commonly present on the surface of old feeding sites left by the raspberry cane midge (*Resseliella theobaldi* (Barnes)).

Control

Spur blight can be minimized in raspberries by rigorous pruning to control the number of canes in a row, so that

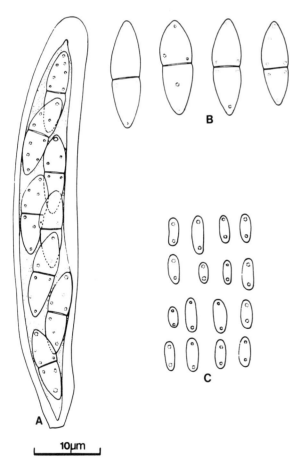

Fig. 6. Ascus containing ascospores (**A** and **B**) and conidia (**C**) of *Didymella applanata*. (Reprinted, by permission, from Punithalingam, 1982)

the overall row width does not exceed 35 cm at the base. Any practice that opens the plant canopy to ensure rapid drying of leaves and canes after rain is beneficial for control and substantially improves spraying efficiency. Similarly, it is important to control weeds and remove cane suckers that arise between and within rows.

Removal of the first flush of young canes with a desiccant herbicide or by cutting or burning has reduced the severity of spur blight in Scotland and Poland. This reduction probably occurs because leaves on replacement canes remain in a resistant juvenile state longer than those of first-flush primocanes. This management technique can only be used in plantings with excessive vigor and with cultivars that are known to tolerate cane removal; otherwise, serious yield losses may result. Because *D. applanata* is specific to *Rubus* and survives the winter only on canes, complete removal of all floricanes greatly reduces or eliminates overwintering inoculum. The use of biennial cropping (alternate-year fruiting) has controlled severe spur blight in New York. Similarly, disease incidence is greatly reduced in fall-bearing red raspberry cultivars if all overwintering canes are removed and burned.

Several contact and systemic fungicides have been effectively used to control spur blight. In the United States, an application of lime sulfur prior to spring growth is widely recommended. Additional sprays used primarily for the control of Botrytis fruit rot are also beneficial. In Scotland, dichlofluanid and thiophanate-methyl applied before harvest to control gray mold have adequately controlled spur blight when sprayed to cover the primocanes as well as the flowers, but they have not increased yields. This may be because the disease was confined to the lower portion of the canes (less than 60 cm above the ground), which contributes little to yield. In such cases, the healthy terminal portions of the cane may compensate by producing additional laterals and fruits. Although only the lower half is generally affected, in exceptionally tall canes this area becomes the cropping zone in the fruiting year after they are cut back in the fall; therefore, its yield potential is important. The position of lesions on the canes in relation to the height at which the canes are headed is important in all assessments of yield loss and efficacy of control programs for spur blight.

Field resistance to *D. applanata* has been reported in the North American cultivars Haida, Viking, Chief, Boyne, Carnival, Chilliwak, and Festival. Inoculation experiments have identified strong sources of resistance for breeding purposes in *R. pileatus* Focke, *R. occidentalis* L., and *R. coreanus* Miq. and in hybrids of these species with red raspberry. Other species showing field resistance that can be used in breeding are *R. crataegifolius* Bunge, *R. lasiocarpus* Sm. (*R. niveus* Thunb.), and wild accessions of *R. idaeus* L. subsp. *strigosus* (Michx.) Maxim. and *R. idaeus* subsp. *vulgatus* Arrhen. The cultivars Leo and Glen Moy, which carry gene *H* (denoting cane pubescence), show strong resistance to *D. applanata*. In addition, the cultivars Glen Clova and Willamette have been shown to be highly tolerant to spur blight, and Meeker has been shown to be very resistant.

Selected References

Blake, C. M. 1980. Development of perithecia and pycnidia of *Didymella applanata* (causing spur blight) on raspberry canes. Trans. Br. Mycol. Soc. 74:101-105.

Burchill, R. T., and Beever, D. J. 1975. Seasonal fluctuations in ascospore concentrations of *Didymella applanata* in relation to raspberry spur blight incidence. Ann. Appl. Biol. 81:299-304.

Goode, J. E. 1970. Nitrogen nutrition and susceptibility of Malling Jewel raspberries to infection by spur blight (*Didymella applanata*). Plant Pathol. 19:108-110.

Koch, L. W. 1931. Spur blight of raspberries in Ontario caused by *Didymella applanata*. Phytopathology 21:247-287.

Pepin, H. S., Williamson, B., and Topham, P. B. 1985. The influence of cultivar and isolate on the susceptibility of red raspberry canes to *Didymella applanata*. Ann. Appl. Biol. 106:335-337.

Punithalingam, E. 1982. *Didymella applanata*. Descriptions of Pathogenic Fungi and Bacteria, No. 735. Commonwealth Mycological Institute, Kew, Surrey, England.

Williamson, B. 1984. Polyderm, a barrier to infection of red raspberry buds by *Didymella applanata* and *Botrytis cinerea*. Ann. Bot. 53:83-89.

Williamson, B., and Dale, A. 1983. Effects of spur blight (*Didymella applanata*) and premature defoliation on axillary buds and lateral shoots of red raspberry. Ann. Appl. Biol. 103:401-409.

Williamson, B., and Hargreaves, A. J. 1981. Effects of *Didymella applanata* and *Botrytis cinerea* on axillary buds, lateral shoots and yield of red raspberry. Ann. Appl. Biol. 97:55-64.

Williamson, B., and Pepin, H. S. 1987. The effect of temperature on the response of canes of red raspberry cv. Malling Jewel to infection by *Didymella applanata*. Ann. Appl. Biol. 101:295-302.

(Prepared by B. Williamson)

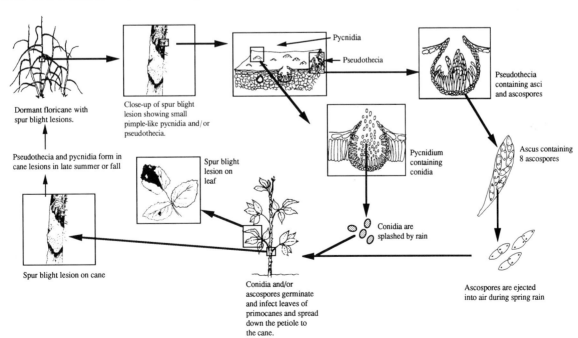

Fig. 7. Disease cycle of spur blight, caused by *Didymella applanata*. The fungus overwinters as pseudothecia and pycnidia in lesions on floricanes that were infected during the previous season. Pseudothecia produce ascospores, and pycnidia produce conidia that serve as a primary inoculum in the spring. (Courtesy M. A. Ellis; drawing by Cindy Gray)

Cane Botrytis

Cane Botrytis, or gray mold wilt, occurs on most *Rubus* cane fruits, but red raspberry is the most seriously affected. This important cane disease is caused by the same fungus (*Botrytis cinerea* Pers.:Fr.) that causes gray mold on fruits of *Rubus* spp. Cane Botrytis has many features in common with spur blight, caused by the fungus *Didymella applanata* (Niessl.) Sacc., and the two diseases are often confused. It is common to find both diseases in the same planting or on the same canes. Cane Botrytis is generally considered to be the more damaging of the two diseases. It was first reported in England in 1931, but isolation of the pathogen and inoculation studies were initially conducted in Nova Scotia in 1952.

Symptoms

Symptoms first appear in mid- to late summer as pale brown lesions on mature to senescent leaves of primocanes. Young or immature leaves are not susceptible. In time, the lesions spread through the petioles to the nodes, and infected leaflets are usually shed prematurely. The fungus can cause a lesion that girdles and kills the primocane (Plate 10); however, this is relatively rare. More typically, pale brown lesions develop and spread rapidly around the cane (Plate 11). Spur blight lesions can be distinguished from those of cane Botrytis by their much darker chestnut brown color. *Botrytis* lesions often show conspicuous banding patterns or watermark symptoms caused by fluctuating rates in the growth of the pathogen, probably related to diurnal temperature changes. One of these lesions from a single infected leaf can spread over three or four internodes. Where two lesions merge, a sharp boundary is always visible; one is also formed where a *Botrytis* and a spur blight lesion converge.

As primocanes turn brown in the fall, *Botrytis* lesions become indistinguishable from healthy tissues. After several weeks of low temperatures the lesions turn gray or whitish and become conspicuous again. Whitening may affect only a portion of a lesion. On some canes the watermark symptom may reappear as white and gray bands (Fig. 8). Prominent sclerotia form beneath the epidermis during winter and erupt in spring as shiny, black, blisterlike structures (Plate 12). During periods of high humidity, these sclerotia become covered with the gray mycelium and conidia of the fungus (Plate 27); they sometimes appear in late summer on lesions derived from early-season infections.

When surface tissues are removed from the cane lesions by scraping, healthy green vascular tissue is revealed. The buds at infected nodes, as with spur blight, are smaller and less likely to produce lateral shoots in spring than those at uninfected nodes.

Causal Organism

The disease is caused by the fungus *Botryotinia fuckeliana* (de Bary) Whetzel, but the anamorph *Botrytis cinerea* is the only state recorded on *Rubus* spp. The fungus is a common pathogen of numerous temperate crop and weed species and is ubiquitous on plant debris. The teleomorph may be important for the maintenance of heterogeneity by sexual recombination but is apparently rare in the field. (For a description of conidiophores and conidia, see Botrytis Fruit Rot [Gray Mold] and Blossom Blight.) Sclerotia are 2–4 mm long and consist of a heavily melanized rind, thin cortex, and large central medulla. They develop in the primary cortex of the cane and lift the epidermis substantially. Although five other *Botrytis* spp. have been shown to produce lesions on wound-inoculated primocanes, only *B. cinerea* has developed sclerotia.

Disease Cycle and Epidemiology

The principal source of inoculum in early spring is probably sclerotia on canes, but most plant debris is colonized by the fungus, which can produce conidia in wet weather. During periods of high humidity, conidia are profusely produced from sclerotia; production is initiated around March and continues throughout the growing season. They are dispersed predominantly by wind, but also by rain-splash. In raspberry plantings, the greatest numbers of conidia have been trapped from the air after the first fruit has ripened. The first lesions on primocanes often appear in July on nodes near the base; as the season progresses, they develop at nodes higher on the canes. Inoculation experiments in Scotland have shown that the infection of leaves and petioles substantially delays the growth of axillary buds, even if the infection occurs in late August. Because only mature to senescent leaves can be infected, the disease is the most severe inside a dense canopy, generally on the lower half of primocanes. In Scotland, most severe outbreaks of cane Botrytis can be found in high-density nursery plantings, but because such plants are dug, planted, and cut back before bud failure occurs in the fruiting year, the disease does not usually result in economic loss.

A primary protective layer of suberized cells prevents the fungus from entering the node on the upper side of the petiole and the mycelium from spreading to the axillary buds. No protective layer is formed on the underside of the petiole; therefore, the lesion spreads unimpeded into the primary cortex of the cane, outside the cork layers (the polyderm). Lesions spread more rapidly in mid- to late August than in July. In Scotland, buds at infected nodes are smaller (and a proportion of them less likely to produce vigorous and productive lateral shoots the following spring) than those at comparable uninfected nodes. The death of axillary buds at infected nodes has been reported on raspberry canes after severe winters in Poland.

If a high proportion of nodes in the upper cropping region of canes is affected by cane Botrytis, bud failure and consequent yield losses may be substantial. And since it is common for spur blight and cane Botrytis to infect the same cane—both fungi occupy the same ecological niche and have similar effects on plant growth—it is difficult to attribute yield loss to only one or the other of these diseases. Additional studies are needed to determine and define the yield loss potential of the two diseases.

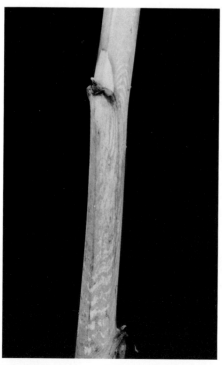

Fig. 8. Watermark symptom on *Botrytis* cane lesion. Note zonate markings. (Courtesy B. Williamson; copyright Scottish Crop Research Institute. Used by permission)

Control

To control cane Botrytis, like most cane diseases, it is important to maintain an open plant habit by pruning to a narrow crop row, to keep the plantation weed-free, and to avoid excessive use of nitrogen fertilizer. These practices allow a rapid drying of foliage after rain and prevent premature senescence from heavy shading, both of which predispose the plant to infection by *B. cinerea*.

The removal of the first flush of primocanes in various cane-vigor-control programs has reduced the severity of cane Botrytis, probably because second-flush canes are less susceptible to infection for a longer period in late summer. Biennial cropping (alternate-year bearing) does not control this disease, probably because inoculum is present on other hosts and crop debris.

Fungicides applied before harvest for the control of gray mold give adequate control if the entire plant is thoroughly covered. Additional applications during and after harvest provide additional protection and may be justified during mild, wet growing seasons. Strains of *B. cinerea* resistant to MBC-generating and dicarboximide fungicides are common in many areas.

The cultivars Chilcotin and Meeker are strongly resistant, and red raspberry derivatives of *R. pileatus* Focke, *R. occidentalis* L., and *R. crataegifolius* Bunge have also been shown by wound inoculation to be highly resistant to cane Botrytis. A higher proportion of laterals at infected nodes emerge in the cultivar Glen Clova than in Malling Orion and Malling Jewel, indicating that some tolerance to this fungus may occur.

Selected References

Bielenin, A., Profic-Alwasiak, H., and Cichocki, J. 1980. Effectiveness of new fungicides in control of raspberry diseases. (In Polish) Pr. Inst. Sadow. Kwiaciarstwa 22:147-153.

Ellis, M. B., and Waller, J. M. 1974. *Sclerotinia fuckeliana*. Descriptions of Pathogenic Fungi and Bacteria, No. 431. Commonwealth Mycological Institute, Kew, Surrey, England.

Harrison, J. G., and Williamson, B. 1986. *Botrytis* spp. on red raspberry: Survival in fruits and infection of canes. Trans. Br. Mycol. Soc. 86:171-173.

Hockey, J. F. 1952. Grey mould wilt of raspberry. Sci. Agric. 32:150-152.

Jennings, D. L., and Williamson, B. 1982. Resistance to *Botrytis cinerea* in canes of *Rubus idaeus* and some related species. Ann. Appl. Biol. 100:357-381.

Labruyère, R. E., and Engels, G. M. M. T. 1963. Fungi as the cause of diseases of the raspberry cane and their connection with the occurrence of the raspberry cane midge. (In Dutch) Neth. J. Plant Pathol. 69:235-257.

Swait, A. A. J. 1980. Field observations on disease susceptibility, yield and agronomic character of some new raspberry cultivars. J. Hortic. Sci. 55:133-137.

Swait, A. A. J. 1982. Fungicide programmes for the control of spur blight, powdery mildew, and grey mould on raspberry. Ann. Appl. Biol. 100:289-295.

Williamson, B., and Hargreaves, A. J. 1981. Effects of *Didymella applanata* and *Botrytis cinerea* on axillary buds, lateral shoots and yield of red raspberry. Ann. Appl. Biol. 97:55-64.

Williamson, B., and Jennings, D. L. 1986. Common resistance in red raspberry to *Botrytis cinerea* and *Didymella applanata*, two pathogens occupying the same ecological niche. Ann. Appl. Biol. 109:581-593.

(Prepared by B. Williamson)

Purple Blotch

Purple blotch is a disease of stems and lateral shoots of blackberries in Europe. The causal fungus was first recognized in France, but most information about the disease has been obtained from inoculation experiments and epidemiological studies in Switzerland, where crop losses of 80% have been reported. The disease also causes yield losses in the Netherlands and southeastern England.

Symptoms

In late summer minute, dark green lesions first appear on the vegetative canes near ground level. As the disease progresses, new lesions appear higher on the canes throughout the winter. They eventually become reddish and then brown, with a conspicuous red margin (Plate 13). In spring their centers become a paler brown and less conspicuous. From late February until April the lesions rapidly expand up to 2 cm in length. In time, they merge and may completely cover stem internodes up to 2.5-3.0 m from the crown.

In early spring, rows of tiny black pycnidia break through the epidermis. Under high humidity, strands of conidia in white mucilage (cirrhi) up to 1 mm long are extruded from the pycnidia and are visible to the naked eye.

In spring, the effects of this disease can be confused with late frost damage. Axillary buds in infected portions of canes often begin to grow normally in spring, but later the flower buds and leaves, initially at the ends of lateral shoots, stop normal development and die. Under severe conditions for disease development the entire shoot appears barren and desiccated.

Disease Cycle and Epidemiology

The pycnidia mature and erupt through the epidermis in early March. Studies in Switzerland have shown that conidia are dispersed by rain from April to August, with peak dispersal in June. This pattern has been confirmed in the Netherlands. Inoculation tests have shown that under conditions favorable for infection, the minimum incubation period is 3 months. Field observations indicate that the period between the peak dispersal of conidia and the appearance of lesions is usually 5-8 months. Conidia germinate at 0-33°C, with the optimum at 21°C. Infection occurs readily through stomata in the undamaged epidermis of young canes. Inoculation tests have shown that no infection of second-year stems occurs.

Experiments in the Netherlands have shown that chilling inoculated stems for 6 weeks at 4°C under light induces lesion development and early maturation of pycnidia.

Causal Organism

Purple blotch is caused by the fungus *Septocyta ruborum* (Lib.) Petr. It has no known teleomorph and has been known under numerous synonyms, of which *Rhabdospora ramealis* (Roberge ex Desmaz.) Sacc. is the most common. The pycnidia (up to 500 μm in diameter) are immersed in the bark. The pycnidial wall has a thick, heavily pigmented outer layer and a hyaline inner layer. The hyaline conidia (20-30 × 1-1.5 μm) are straight or sometimes curved and one- to three-septate.

Control

New plantations should be established, using healthy plants grown in an area free of wild blackberries. Infected canes should be removed and burned.

In Switzerland, two sprays of copper-containing fungicide, applied in late April and mid-June, have given satisfactory control. Sprays of benomyl and copper have been used in England to control *S. ruborum*, but isolates resistant to MBC-generating and copper fungicides have developed in some plantings.

Selected References

Koellreuter, J. 1950. Morphology and biology of *Rhabdospora ramealis* (Desm. & Rob.) Sacc. (In German) Phytopathol. Z. 17:129-160.

Oort, A. J. P. 1952. Die-back of blackberry caused by *Septocyta ramealis* (Rob.) Pet. (In Dutch) Tijdschr. Plantenziekten 58:247-250.

Punithalingam, E. 1980. *Septocyta ruborum*. Descriptions of Pathogenic Fungi and Bacteria, No. 667. Commonwealth Mycological Institute, Kew, Surrey, England.

(Prepared by B. Williamson)

Ascospora Dieback

Occasional outbreaks of Ascospora dieback have occurred in North America and Europe on raspberries and blackberries. The pathogen also occurs on a wide range of woody hosts and other plants. Its effect on host growth, yield, and fruit quality in *Rubus* spp. is not known, and the importance of this disease is not clearly understood. Present evidence suggests that Ascospora dieback requires host predisposition for infection to occur, and that the fungus may infect only after low-temperature injury.

Symptoms

Lesions first appear in late summer or early autumn on primocanes of red and black raspberries but are most marked on fruiting canes of blackberries, particularly after winter injury. Lesions are ashen white on red raspberries and blackberries and often have a reddish margin (Plate 14). On black raspberry they are bluish with a silvery bloom. Affected areas are usually 7–20 cm long and rarely envelop the stem. In the early stages of infection, the discoloration of host tissue extends only a few cells below the epidermis. In early spring, the lesions become dotted with reddish brown acervuli. Occasionally the production of acervuli and sporulation may commence by late August in the previous year. After conidia have been discharged, the bark surrounding the lesion becomes sooty black, and the epidermis may split and peel back in affected areas. Lesions develop primarily at nodes, indicating that infection may occur through petioles or leaf scars. However, the inoculation of wounded petioles in Scotland produced no nodal lesions on a range of *Rubus* genotypes. Symptoms were reproduced in Scotland by inoculating wounded stem internodes of red and black raspberries, blackberries, and blackberry–red raspberry hybrids with mycelium in July, but no vascular damage or other injury occurred.

Causal Organism

Ascospora dieback is caused by the fungus *Clethridium corticola* (Fuckel) Shoemaker & E. Müller (anamorph *Seimatosporium lichenicola* (Corda) Shoemaker & E. Müller). Ascocarps are produced on dead raspberry canes at least 2 years old. Ascocarps are solitary and subcuticular, with a short, projecting beak, and measure 200–300 µm high and 300–500 µm wide. Their walls are two to five layers thick, but are thicker near the ostiole, which is lined by hyaline periphyses 2 µm wide. Paraphyses are absent. The eight-spored asci (100–200 × 8–10 µm) form in a broad hymenium. Asci are cylindrical and thin-walled, except at the apex, which has an iodine-positive ring and a pulvillus, which stains with india ink. The ascospores are elliptical, rounded at the ends, and constricted at the first-formed septum, often with three close septa and long end cells (sometimes with five septa and, rarely, with a longitudinal septum in one cell). The ascospores are hyaline and 11–17 × 5–7 µm; five-septate spores are as large as 15–18 × 6–7 µm. The black conidiomata (acervuli) are peridermal to subperidermal and up to 250 µm in diameter. Conidiophores are branched and septate from the base, hyaline, and up to 27 µm long × 1.5–2.5 µm wide. The conidia (Fig. 9) are formed from percurrently proliferating conidiogenous cells (annelides); they are straight, fusiform to clavate, smooth, three-euseptate, and 13–15 × 5.5–6.5 µm, with periclinal walls often collapsed and the two median or three upper cells darker brown than the end cells.

Both the teleomorph (ascomycete) and the anamorph (coelomycete) have been known under various names. The name *Ascospora dieback* is a misnomer. The teleomorph was once placed in the genus *Ascospora* Fr. as *A. ruborum* (Oudem.) Zeller. This generic name is now considered a synonym of *Omphalospora* Theiss. & Syd., a genus in the Dothideaceae, which is not the correct taxonomic place for this pathogen. *A. ruborum* is a synonym of *C. corticola*. There are several other names for the ascomycete state, which correctly belongs in the Amphisphaeriaceae. The coelomycete state has even more synonyms. It has been ascribed to *Coryneum ruborum* Oudem. by Zeller, to *Hendersonia rubi* Sacc. by Moore, and to *Coryneopsis* Grove by Moore. None of these genera are appropriate for the organism. *Coryneopsis* is a synonym of *Seimatosporium*.

Control

No control methods for this disease have been reported.

Selected References

Brockmann, I. 1975. Untersuchungen über die Gattung *Discostroma* Clements (Ascomycetes). Sydowia 28:275-338.

Ferrata, M., and D'Ambra, V. 1981. Studio di un'alterazione del lampone causata da *Seimatosporium lichenicola* (CDA) Shoemaker e Muller (*Coryneum ruborum* Oud.). Riv. Patol. Veg. Ser. IV 17:15-22.

Moore, W. C. 1943. Diseases of crop plants. A 10 year's review (1933-1942). Bull. Minist. Agric. (London) 126:1—100.

Moore, W. C. 1959. British Parasitic Fungi. Cambridge University Press. 430 pp.

Shoemaker, R. A., and Müller, E. 1964. Generic correlations and concepts: *Clathridium* (= *Griphosphaeria*) and *Seimatosporium* (= *Sporocadus*). Can. J. Bot. 42:403-410.

Sutton, B. C. 1975. Coelomycetes V. *Coryneum*. Mycol. Pap. 138:1-224.

von Hohnel, F. 1918. Mykologische Fragmente. Ann. Mycol. 16:35-174.

Zeller, S. M. 1925. *Coryneum ruborum* Oud. and its ascogenous stage. Mycologia 17:33-41.

(Prepared by B. C. Sutton and B. Williamson)

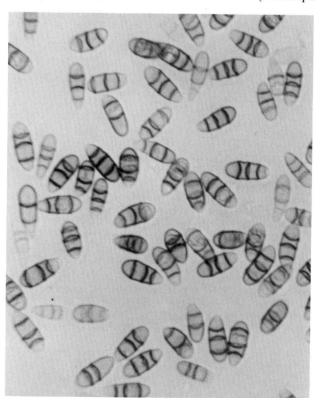

Fig. 9. Conidia of *Clethridium corticola* (Ascospora dieback). (Courtesy B. Williamson; copyright Scottish Crop Research Institute. Used by permission)

Botryosphaeria Cane Canker of Blackberry

One of the more serious cane canker diseases of thornless blackberry plants (*Rubus* subg. *Eubatus* spp. hybrids) is caused by *Botryosphaeria dothidea* (Moug.:Fr.) Ces. & De Not., which occurs in the eastern United States. This cane canker disease is highly destructive, often killing canes and reducing fruit yields to uneconomic levels. In some cases, particularly susceptible cultivars have had to be replaced with more resistant ones or entire plantings abandoned. The disease has been observed on both wild and cultivated thorny blackberries; however, damage to canes is sporadic, and the economic importance is apparently minimal.

Symptoms

Symptoms caused by *B. dothidea* are similar to those of spur blight on red raspberry canes. Cankers generally develop around one or more buds on the main stem of the second-year floricanes and appear as reddish to dark reddish brown discolorations below or to one side of the subtending leaf petiole (Plate 15). The bud or lateral shoot at the infected node is usually killed. Occasional cankers develop at wound sites along internodes. Lesions develop around and beneath the node and later spread above it, girdling the cane. The disease is usually not evident until the fruit ripens. At that time leaves in cankered portions of the canes begin to wilt, turn yellow, and desiccate, but entire canes may be affected. Generally, fruit development ceases on affected canes. Developing cankers often have a zonate pattern of dark and lighter coloration, and old ones become light-colored (to silvery gray) on dead canes. Well-developed cankers often split open longitudinally, exposing the pith. Pycnidia of the fungus are scattered subepidermally throughout the canker, becoming erumpent and exposing their ostioles.

Symptoms similar to Botryosphaeria cane canker may also be caused by other fungi, winter freeze damage, or crown borer infestations. Because several fungi cause cankers that are similar in appearance, identification in the field may be difficult. In addition, two or more of these pathogens may occur on the same stems or cause confluent cankers.

Causal Organism

The fungus is known to be the anamorph of *B. dothidea*. No acceptable name for the conidial state is currently recognized. The fungus also causes Botryosphaeria fruit rot and canker of apples as well as diseases of many other woody plants. The teleomorph has not been observed on blackberry or in cultures of blackberry isolates.

Pycnidia (52–65 μm in diameter) on blackberry stems are immersed to erumpent, globose to irregularly globose, and black, with well-defined ostioles and elongate necks (35–37 μm long). Conidiogenous cells line the cavity of the pycnidium. Conidiophores are short, hyaline, and unbranched and arise from the cell layer lining the interior of the pycnidium. Conidiophores are holoblastic phialides, and portions of the outer conidiophore wall may be observed at the truncate ends of conidia. Conidia (6.3–8 × 16.5–20 μm) are hyaline, single-celled, ellipsoid to fusoid, distinctly basally truncate, and multinucleate (as many as eight to 10 nuclei per conidium). Conidia often become two-septate (occasionally one- or three-septate), with the middle cell often darker than the others.

Colonies of the anamorph of *B. dothidea* on potato-dextrose agar grow rapidly, covering the entire surface of the medium in a 10-cm petri plate within 4 days at 20–25°C. Initial colonies are white and cottony and turn dark gray and floccose after 3–5 days. Numerous solitary black conidiomata (200–350 μm in diameter) develop on the medium surface. No color is imparted to the growth medium by the fungus.

Disease Cycle and Epidemiology

The epidemiology of this disease has not been fully determined, but buds with persistent leaf petioles are apparently the primary site of infection. However, artificial inoculation of wounded and unwounded nodes and internodes of canes and the occurrence of natural infections around wounds in internodal areas indicate that wounds are also important sites of entry for the fungus. *B. dothidea* is generally considered to be a wound-invading pathogen on other hosts.

Blackberry canes wounded by pruning in the summer or winter are often colonization sites for several fungi, but only infrequently for *B. dothidea*. It is not clear if the wounding of buds or damage from winter injury is essential for the natural infection of canes, or if healthy tissues may be attacked directly. Controlled freezing experiments with European white birch (*Betula alba* L.) and European mountain ash (*Sorbus aucuparia* L.) have shown that both species are predisposed to attack by *B. dothidea* after they are injured by freezing. Leaf tissue infection does not appear to be important in the development of the blackberry cane canker disease, as it is with spur blight of red raspberry. Young lesions (less than 2 cm in diameter) on canes have been found throughout the growing season, indicating that infection may take place any time from early spring to fall.

The fungus can overwinter in cankers in dead canes and probably also in cankers on living primocanes. In addition, *B. dothidea* has been isolated from dormant buds, petiole scars, and retained petioles, indicating that the fungus can overwinter in these tissues. In many areas, inoculum may be produced from cankers on other woody hosts, such as apple. Once *B. dothidea* becomes established in a blackberry planting, the biennial growth habit of blackberry plants ensures perpetuation of the fungus.

Once the fungus is established saprophytically in senescent leaf petioles, it can invade adjacent healthy cane and bud tissue. Cultivars with a high incidence of disease often tend to retain a high percentage of petioles into midwinter.

Control

Disease-free nursery stock should always be used to begin new plantings. Since *B. dothidea* is a common pathogen of other fruit crops, such as apple, proximity to these potential sources of inoculum should be considered when establishing new plantings.

Plantings should be fertilized to maintain plant vigor, but the use of excessive fertilizer, especially nitrogen, should be avoided. Fungicide control measures have not been established for this disease. Some resistance in thornless blackberry cultivars has been noted in field plot studies. However, greenhouse inoculation tests showed that none of the cultivars were immune to infection. The cultivars Chester, Thornless, and Dirksen Thornless are more resistant to Botryosphaeria cane canker than Black Satin and Thornfree, which are of intermediate susceptibility; Hull Thornless and Smoothstem are still more susceptible.

Selected References

Maas, J. L. 1986. Epidemiology of the *Botryosphaeria dothidea* cane canker disease of thornless blackberry. Acta Hortic. 183:125-130.

Maas, J. L., and Uecker, F. A. 1984. *Botryosphaeria dothidea* cane canker disease of thornless blackberry. Plant Dis. 68:720-726.

Wine, E. G., and Schoeneweiss, D. F. 1980. Localized freezing predisposition to *Botryosphaeria canker* in differentially frozen woody stems. Can. J. Bot. 58:1455-1458.

(Prepared by J. L. Maas)

Rosette (Double Blossom)

Rosette, or double blossom, is a serious disease of most cultivars of erect blackberries and is limited to the genus *Rubus*. It occurs from New Jersey to Illinois and southwest to Texas and is considered a major disease of blackberries in the southeastern United States. It has not been reported outside the United States or from the Pacific Coast. Rosette also occurs on trailing blackberries and dewberries, but rarely on red and black raspberries. Losses from rosette are reduced yield, poor-quality fruit, and (in severe cases) the death of canes.

Symptoms

The symptoms of rosette are usually striking and can result in a complete change in the physical appearance of the plant (Plate 16). Buds on blackberry primocanes usually become infected in early summer. There are no symptoms until the following spring, when several leafy shoots develop from each infected vegetative bud (Plate 17). These shoots are generally smaller than normal and have pale green foliage that later turns bronze. The proliferation of shoots is commonly referred to as rosette, or *witches'-broom*. One or all buds on an infected cane may produce rosettes.

Unopened infected flower buds are usually elongated and larger, coarser, and frequently redder than uninfected buds. Sepals on infected flowers enlarge and occasionally differentiate into leaves (Plate 18). As the petals unfold, they are usually pinkish, wrinkled, and twisted, which gives the appearance of double flowers; this is why the disease is commonly called *double blossom*. Occasionally they become green and leaflike. The pistils are usually larger and longer than normal and sometimes become abnormally shaped. Often the pistils and anthers are brown. The fungal mycelium grows over the surface of infected pistils and stamens and produces a whitish spore mass. Berries do not develop from infected blossoms, and uninfected parts of the same cane often produce only small, poor-quality fruit. In some cultivars the rosette may fail to develop, but infected blossoms are usually sterile.

Causal Organism

Rosette is caused by the fungus *Cercosporella rubi* (G. Wint.) Plakidas. Its mycelium is hyaline and septate, with hyphae 1.5 μm or less in diameter. Aerial hyphae are 2.0–2.5 μm in diameter and anastomose freely to form a complex network. Conidiophores arise as side branches, either singly or in fascicles. They are hyaline, short, and average 3.8×12.7 μm. They are usually unbranched, lack septa, and have one to seven spore scars. Conidia are hyaline, cylindrical to tapering, straight to curved, and zero- to 12-septate (average three-septate). Conidia may or may not have constrictions at septa and are quite variable in length, ranging from $13–96 \times 2.7–4.7$ μm (average 33.8×3.8 μm). The mycelial stage occurs within vegetative and floral buds of the host, and conidia are formed in open blossoms. The mycelium may be observed with a microscope in meristematic tips of shoots from rosettes when the tips are freshly mounted in cotton blue/lactophenol. The fungus can be readily isolated from infected buds just prior to opening, and it forms a slow-growing culture, which has a rough, slippery, and rubbery consistency on potato-dextrose agar.

Disease Cycle and Epidemiology

The disease cycle follows the biennial habit of blackberry canes. As the primocanes emerge in the spring, the fungus sporulates on infected flowers of floricanes (Fig. 10). Conidia are disseminated to the young primocanes, where they germinate and infect young axillary buds. The fungus then grows between the bud scales surrounding the embryonic bud tissues. As secondary (accessory) buds develop beside an infected one, they also become infected. Infected buds usually remain symptomless until the next spring, but a few may "break bud" in an unusually warm and late fall. The fungus overwinters in infected buds, and during the winter bud proliferation is induced. When infected buds break dormancy in the spring, they develop numerous short, abnormal, and off-colored lateral shoots, which form witches'-broom. The fungus physiologically alters the bud so that it produces an indeterminate vegetative branch with sterile flowers rather than a determinate fruiting lateral. Affected lateral shoots are the first to grow and produce abnormal blossoms, upon which the fungus sporulates heavily on stamens and pistils. Conidia are dispersed by wind and insects to the newly formed axillary buds on primocanes. The fungus infects these buds and overwinters in them, thus completing the disease cycle. Infection occurs in the spring and summer (well into harvest), but may be limited by temperatures above 30°C. In erect blackberries the fungus is confined to the buds and has not been found in other tissues. In some trailing blackberry cultivars propagated by tip rooting, the fungus can become established in the crown of a plant produced from the rooted tip of a rosetted cane. Buds on the primocanes arising from infected crowns have also been infected.

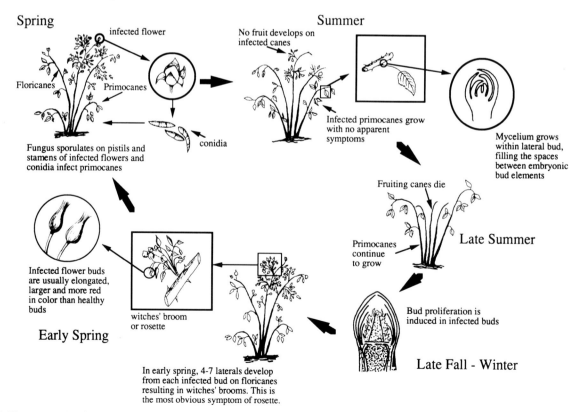

Fig. 10. Disease cycle of rosette, or double blossom. (Drawing by Cindy Gray)

Control

The control of rosette should begin with the selection of a site isolated from wild blackberries and dewberries. Disease-free nursery stock should always be used to begin new plantings, and root cuttings, rather than mature rooted plants, should be used, because the fungus does not occur in roots.

In areas where the disease is not severe, rosette can be controlled by sanitation. Infected rosettes and blossom clusters should be handpicked and removed before they open, to prevent dispersal of the fungus. In addition, old floricanes should be removed and destroyed immediately after harvest. The removal of all wild blackberries and dewberries in and around the planting is recommended.

In areas where the disease is severe, rosette can be controlled with or without the use of fungicides, depending upon the vigor of the blackberry cultivar and the length of the growing season after harvest. For vigorous cultivars in a location where adequate growth occurs after harvest, primocanes as well as floricanes may be cut off to ground level immediately after harvest. This practice removes infected tissue and inoculum. The primocanes are then allowed to regrow from buds at the base of the plant. This technique is effective in controlling the disease but may reduce yields. In locations where yields obtained from using this method are inadequate, control may be achieved by cutting the floricanes off at ground level in the spring before primocanes emerge. This biennial cropping technique requires two plantings, which are cut in alternate years, to provide a crop on one of the plantings each year. Careful attention to the exclusion of inoculum from all outside sources must be emphasized when using this control method.

The chemical control of rosette depends on timing fungicide applications to coincide with infection periods during the blooming of the rosettes. This period varies with location and cultivar. It begins at first bloom and may extend past harvest at some locations. Bordeaux mixture and benomyl have provided effective control when the applications were properly timed.

Selected References

Cook, M. T. 1911. The double blossom of the dewberry. Del. Agric. Exp. Stn. Bull. 93. 12 pp.

Moore, J. N. 1980. Blackberry production and cultivar situation in North America. Fruit Var. J. 34:36-41.

Moore, J. N., Bowden, H., and Slack, D. A. 1977. Chemical control of rosette in blackberries: Preliminary results. Arkansas Farm Res. 26(2):10.

Plakidas, A. G. 1937. The rosette disease of blackberries and dewberries. J. Agric. Res. 54:275-303.

(Prepared by B. J. Smith and J. A. Fox)

Downy Mildew

Downy mildew of *Rubus* spp. occurs in production areas worldwide (except South America), but is generally rare. In New Zealand the disease is extremely important on boysenberry. More recently, serious outbreaks of the disease on boysenberry, other blackberry–red raspberry hybrids, and blackberry have occurred in California, Washington State, and in the United Kingdom. The disease can be found on the leaves, petioles, primocanes, pedicels, calyces, and fruits of blackberry and red raspberry cultivars, and particularly on the hybrids boysenberry, youngberry, ollalieberry, and tummelberry. On wild European blackberries and red raspberry the pathogen is restricted mainly to the leaves. On highly susceptible blackberry–red raspberry hybrids, and under favorable conditions, the fungus spreads systemically through most of the plant.

Under conditions conducive to disease development, severe losses occur as a result of pedicel and fruit infection, which cause the berry to rapidly dry and split (hence the name *dryberry* for this disease in New Zealand). Plants seem to be particularly susceptible to infection during the propagation of nursery stock. Recently, severe losses from downy mildew have occurred in *Rubus* nurseries in Washington State and the United Kingdom.

Symptoms

Downy mildew on leaves first appears as a yellow discoloration on the upper surface, which soon changes to carmine red to purple. Lesions are usually angular and restricted by veins (Plate 19). In some cultivars, leaf symptoms appear as distinctive purple blotches or lesions (Plate 20), which generally extend along the midrib and lateral veins. On the lower leaf surface, light pink to tan areas appear directly below the blotches on the upper surface. Spore masses are produced only on the lower leaf surface and are initially white but become light gray with age.

Systemic infection of unfolding leaves results in a mosaic pattern of small yellow and red angular lesions, with severe distortion in addition to the red discoloration. As infected leaves age, bright yellow margins often develop around lesions with dead, brown centers. Severely infected leaves often fall prematurely. In New Zealand, symptoms of leaf yellowing and dieback without the blotchiness or reddening occur on loganberry. Foliar infections that result from direct penetration of the fungus may result in scattered leaf spots not bounded by veins.

Suckers arising from systemically infected plants often show some stunting, and terminal leaves often develop a reddish discoloration. Infected canes and fruiting laterals also develop reddened irregular blotches on the epidermis.

The infection of fruit results in a loss of sheen, and the berries appear dull. Early infection of green fruit induces premature reddening, and the berries shrivel and harden rapidly. Those infected at later stages of development often split into two parts, one or both becoming shriveled (Plate 21). Infected pedicels become dry and often show a pronounced red discoloration on one or more sides.

Causal Organism

The taxonomy of the downy mildew fungus of *Rubus* spp. is confusing. In New Zealand and California the fungus was named *Peronospora sparsa* Berk., based on the presence of sporangia and size of oospores. *P. sparsa* is known to cause downy mildew of roses. In Europe, the pathogen has been identified as *P. rubi* Rabenh., since until recently oospores have not been observed there.

Cross-inoculations performed in vitro with isolates from roses and *Rubus* in the United Kingdom indicate that isolates from either host can infect the other. Therefore, these fungi are probably synonymous, and *P. sparsa* would be the name adopted because of nomenclatural priority.

The pathogen is an obligate parasite. Intercellular hyphae are aseptate and hyaline. The fungus produces filamentous, intercellular haustoria in the mesophyll tissues of leaves and in the outer layers of the cortex parenchyma of leafstalks and canes.

Sporophores of *P. sparsa* are 490–600 × 4–6 μm, with trunks 300–465 μm long. They branch dichotomously three or four times, with branch ends that are 12–16 μm long, slender, and tapered (Fig. 11). One of each branch pair usually curves inward, and one is reflexed. Spores (18–24 × 16–20 μm) are ovoid to ellipsoidal and colorless to pale yellow. A short stalk (pedicel) may occasionally be present on detached spores. Oospores are 22–30 μm in diameter and have a hyaline outer wall; they have not been reported from field specimens in Europe, but were found in detached inoculated leaves in vitro.

Disease Cycle and Epidemiology

Downy mildew occurs most commonly in warm, humid production areas in New Zealand and North America, although in California severe outbreaks have occurred in the arid interior San Joaquin Valley after spring rainfall. The disease is most prevalent during wet weather at temperatures of 18–22°C, which generally occur in mid- to late spring and autumn.

The pathogen overwinters as mycelium inside roots, crowns, and canes. When sucker growth begins in the spring, *P. sparsa* closely follows apical shoot growth, infecting new stems and emerging leaves. Infected leaves on primocanes become the initial sites for sporulation, but leaves on infected fruiting laterals may also produce inoculum. Systemic leaf infection requires wet or humid conditions when leaves are unfolding. Sporulation is usually found in dense foliage near the cane or at ground level, where the humidity is highest. Weed growth and dense canopies favor the production of diseased suckers and fruiting laterals.

Airborne spores are produced during cool, wet nights and are disseminated by wind to foliage, including that of new primocanes, blossoms, and developing berries. In New Zealand, oospores develop in leaves and sepals in the field in early summer.

Both pathogen and disease development are favored by cool to moderate temperatures, high humidity, and wet weather.

Symptoms develop within 10–11 days after infection, and sporulation occurs in an additional 5–11 days at 18°C. The propagation of systemically infected plants has resulted in a disease incidence of 100% in newly planted fields in the United States. The rapid infection of newly weaned plants from micropropagation in the United Kingdom was probably initiated by airborne inoculum rather than systemically infected stocks. The importance of oospores in the disease cycle is unknown, but their presence in infected leaves may provide an inoculum source in propagation houses.

Control

Pathogen-free planting stocks should be used, and sites with a history of the disease should be avoided. Alternate hosts, such as rose or wild blackberries, in close proximity to the planting should be destroyed.

After the planting is established, the chemical removal of suckers helps to limit early inoculum production. In addition, sucker removal and weed control also reduce humidity and the duration of wetness at the base of the plant, thus helping to prevent conditions favorable for infection. Old fruited canes should be removed and destroyed (burned or buried) immediately after harvest to reduce inoculum buildup.

Fungicide sprays should be applied in the spring to protect new foliage, flowers, and developing berries from infection. Systemic fungicides such as metalaxyl, which has been used in commercial *Rubus* plantings in California and New Zealand, provide the best control. Since downy mildew fungi on other crops have developed a resistance to metalaxyl, it is important that such fungicides not be used to treat nursery stocks and newly propagated plants in nurseries; otherwise, the risk of selecting and disseminating strains of *P. sparsa* with resistance to these fungicides will be substantially increased.

Selected References

Francis, S. M. 1981. *Peronospora sparsa*. Descriptions of Pathogenic Fungi and Bacteria, No. 690. Commonwealth Mycological Institute, Kew, Surrey, England.

Hall, G. 1989. *Peronospora rubi*. Descriptions of Pathogenic Fungi and Bacteria, No. 976. Mycopathologia 106:195-197.

Hall, H. K., and Shaw, C. G. 1982. Oospores of *Peronospora sparsa* Berk. on *Rubus* species. N.Z. J. Exp. Agric. 10:429-432.

Hall, H. K., and Shaw, C. G. 1987. Brambles: Downy mildew on wild and cultivated *Rubus* species in New Zealand. N.Z. J. Exp. Agric. 15:57-60.

McKeown, B. 1988. Downy mildew of boysenberry and tummelberry in the UK. Plant Pathol. 37:281-284.

Tate, K. G. 1981. Aetiology of dryberry disease of boysenberry in New Zealand. N.Z. J. Exp. Agric. 9:371-376.

Tate, K. G., and van der Mespel, G. J. 1983. Control of dryberry disease (*Peronospora sparsa*) in boysenberry with fungicides. N.Z. J. Exp. Agric. 11:141-146.

Wallis, W. A., Shattock, R. C., and Williamson, B. 1989. Downy mildew (*Peronospora rubi*) on micropropagated *Rubus*. Acta Hortic. 262:227-230.

Williamson, B., Wallis, W. A., and Shattock, R. C. 1989. Downy mildew of *Rubus* cane fruits. Pages 113-115 in: Rep. Scott. Crop Res. Inst. 1988.

(Prepared by W. D. Gubler)

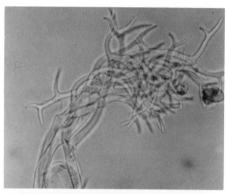
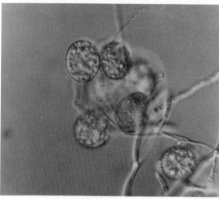
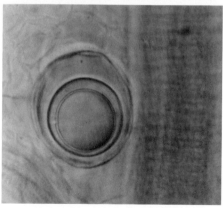

Fig. 11. Sporangiophores (top), sporangia (center), and oospore (bottom) of *Peronospora sparsa*, causal agent of downy mildew. (Courtesy B. Williamson; copyright Scottish Crop Research Institute. Used by permission)

Powdery Mildew

Powdery mildew affects susceptible cultivars of red, black, and purple raspberries, but blackberries and hybrids are usually not affected. The disease can be severe (varying from year to year) on highly susceptible cultivars, and these plants may be stunted and less productive. The infection of flower buds may reduce the quantity of fruit, and infected fruit may be lower in quality or unmarketable as a result of the unsightly covering of mycelial growth.

Symptoms

Infected leaves develop light green blotches on the upper surface. Generally, the lower surface of the leaf directly beneath these spots becomes covered by white, mealy mycelial growth of the powdery mildew fungus. The leaf spots may appear water-soaked. Infected leaves are often mottled, and if surface growth of the fungus is sparse, they often appear to be infected by a mosaic virus. Infected shoot tips may also become covered with mycelial growth (Plate 22). When severely infected, the shoots become long and spindly (rat-tailed), with dwarfed leaves that are often curled upward at the margins (Plate 23). Infected fruit may become covered with a white, mealy, mycelial mat. When the disease is severe, the entire plant may be stunted.

Causal Organism

Powdery mildew on raspberry is caused by *Sphaerotheca macularis* (Wallr.:Fr.) Lind. This organism also attacks strawberry, but cross-inoculations have shown that isolates from strawberry do not infect raspberry, and vice versa. The fungus produces abundant superficial mycelium with haustoria inside host epidermal cells. Conidiophores produce oval, unicellular conidia (25–38 × 15–23 µm) in chains. Reddish brown cleistothecia (58–120 µm in diameter) are formed on host surfaces in the fall. They are covered with long, straight or twisted, dark brown appendages and contain a single eight-spored ascus (45–90 × 50–72 µm), which is elliptical to subglobose (Fig. 12). Ascospores average 22 × 15 µm; they require a winter rest period inside the cleistothecium in order to germinate.

Disease Cycle and Epidemiology

The fungus overwinters as mycelium in buds on shoot tips in Minnesota, but in California it has been reported to overwinter only as cleistothecia, producing ascospores as primary inoculum in the spring. Conidia are generally abundantly produced on the surface of infected tissue, and these serve as secondary inoculum for repeated cycles of infection throughout the growing season. They are airborne and probably remain viable for no more than 21 days. The development of this disease, like most other powdery mildew diseases, is favored by warm, dry weather.

Control

Cultivars differ widely in resistance, and leaf infection is not closely correlated with resistance to fruit infection. Where powdery mildew is a persistent and serious problem, highly susceptible cultivars, such as Viking, Ottawa, Latham, and Glen Clova, should be avoided. Chief, Marcy, and Malling Orion are resistant to the disease, and Logan is immune. *Rubus coreanus* Miq. is a useful donor of resistance in breeding programs.

The removal of late-formed suckers with powdery mildew symptoms and the cutting back of floricanes to a horticulturally desirable height in the spring may help to reduce sources of primary inoculum. Practices that allow good air circulation (such as proper plant spacing, cane thinning, and maintaining narrow rows) have been reported to be helpful in control.

Many of the common fungicides used for control of anthracnose and spur blight, such as captan and ferbam, are not effective for powdery mildew. A delayed-dormant application of lime sulfur was reported to be effective in Oregon, but not in Michigan or Minnesota. In California, sulfur dust has been recommended for powdery mildew control, but it was phytotoxic in other locations.

Benomyl, dichlofluanid, and vinclozolin, when used to control gray mold fruit rot, caused by *Botrytis cinerea* Pers.:Fr., have given some suppression of fruit and foliar mildew; however, under heavy disease pressure, more effective powdery mildew fungicides may be required. Many of the ergosterol biosynthesis-inhibiting fungicides have performed well against powdery mildew. In Scotland triadimefon, bupirimate, and fenarimol (applied in a two-spray program, with sprays applied at the early white bud stage and at 10% flowering) have provided satisfactory control in most seasons. In seasons of high disease pressure, a third spray at full bloom may also be required.

Selected References

Brokenshire, T. 1984. The control of powdery mildew on strawberries and raspberries. Pages 370–375 in: Proc. Crop Prot. North. Brit.

Keep, E. 1968. Inheritance of resistance to powdery mildew, *Sphaerotheca macularis* (Fr.) Jaczewski in the red raspberry, *Rubus idaeus* L. Euphytica 17:417-438.

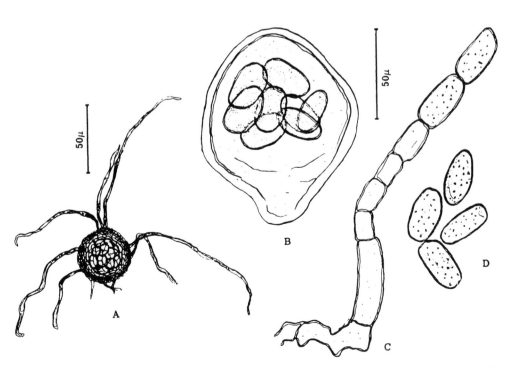

Fig. 12. Cleistothecium (**A**), ascus (**B**), conidiophore (**C**), and conidia (**D**) of *Sphaerotheca macularis*. (Reprinted, by permission, from Mukerji, 1968)

Mukerji, K. G. 1968. *Sphaerotheca macularis*. Descriptions of Pathogenic Fungi and Bacteria, No. 188. Commonwealth Mycological Institute, Kew, Surrey, England.

Peries, O. S. 1962. Studies on strawberry mildew, caused by *Sphaerotheca macularis* (Wallr. ex Fries) Jaczewski. I. Biology of the fungus. Ann. Appl. Biol. 60:211-224.

Peterson, P. D., and Johnson, H. W. 1928. Powdery mildew of raspberry. Phytopathology 18:787-796.

Salmon, E. S. 1900. A monograph of the Erysiphaceae. Mem. Torrey Bot. Club 9:1-292.

Swait, A. A. J. 1982. Fungicide programmes for the control of spur blight, powdery mildew and grey mold on raspberry. Ann. Appl. Biol. 100:289-295.

(Prepared by M. A. Ellis)

Raspberry Leaf Spot

Raspberry leaf spot occurs only on raspberries. A similar disease (Septoria leaf spot) occurs on erect and trailing blackberries. Raspberry leaf spot occurs in most parts of North America but is generally most severe toward the southern limits of the raspberry production region. When severe, it can cause premature defoliation in late summer and early fall, resulting in a reduction of plant vigor and making plants more susceptible to winter injury.

Symptoms

Greenish black, circular to angular spots develop on the upper surfaces of new leaves (Plate 24). As the leaves mature, the spots become whitish or gray, with a well-defined margin, and enlarge to 1-2 mm (sometimes 4-6 mm) in diameter. Leaf lesions sometimes drop out, producing a shot-hole effect. Severely infected leaves drop prematurely, which can lead to serious defoliation in some growing seasons. Inconspicuous, sometimes necrotic cane lesions also develop, particularly toward the bases of the canes.

Causal Organism

The fungus causing raspberry leaf spot is *Sphaerulina rubi* Demaree & M. S. Wilcox. Its taxonomy has been confused with that of the fungus causing Septoria leaf spot on blackberry. Demaree and Wilcox considered the anamorph to be *Cylindrosporium rubi* Ellis & Morg. There is evidence that *Septoria rubi* Westend. var. *brevispora* Sacc. is the same fungus. The latter was elevated to species rank as *Septoria brevispora* by Zeller but is now known as *Septoria darrowii* Zeller, because of prior publication of *S. brevispora* Ellis & J. J. Davis for a different fungus. *S. darrowii* has priority over *C. rubi*. It has conidia that are 32–86 × 3–4.8 μm and three- to nine-septate, unlike the conidia of *Septoria rubi*, which are 40–55 × 1.5 μm and two- to three-septate. In 1937, Zeller presented data showing marked differences in pathogenicity, symptom production, and fungal morphology between collections of infected raspberry leaves from Maryland and Oregon, suggesting a lack of homogeneity in the fungi causing raspberry leaf spot. Most workers agree that the fungi that cause leaf spot on raspberries and erect and trailing blackberries do not cross-infect under field conditions. Throughout most of North America the leaf spot diseases of both raspberry and blackberry are commonly referred to as Septoria leaf spot. In the southern United States a leaf-spotting fungus—appropriately named *Mycosphaerella confusa* F. A. Wolf (anamorph *Cercospora rubi* G. Wint.)—also occurs on raspberries and blackberries.

Perithecia (88–140 × 86–120 μm) of *Sphaerulina rubi* are usually numerous and may be scattered or in groups. They are mostly hypophyllous, innate, erumpent, black, conical, and ostiolate to papillate. Paraphyses are lacking. Asci (44.8–70.0 × 9.6–15 μm) are eight-spored, fasciculate, and sessile and may be curved or straight. The outer wall of the ascus is about 2 μm thick, and the inner wall is membranous. Ascospores (32–5.8 μm) are four- to eight-celled (usually four-celled), hyaline, granular, cylindrical, usually curved, and pointed at both ends. Pycnidia (58–80 × 58–12 μm) are epiphyllous and subepidermal and lack an ostiole. Pycnidial walls are thin and usually one to three cells thick. Conidia (32–86 × 3–4.8 μm) are long and slender, hyaline, elongate, obclavate, slightly curved to falcate, pointed at one end, and three- to nine-septate.

Disease Cycle and Epidemiology

Sphaerulina rubi commonly overwinters as perithecia on dead leaves and as pycnidia on leaves and canes. Waxy masses of conidia from cane lesions are an important source of primary inoculum in early spring. Only young expanding leaves and canes are susceptible to infection by either ascospores or conidia in the spring and early summer. Pycnidia rapidly develop on diseased tissue during the summer. They exude masses of conidia, which are spread by splashing or wind-driven rain and can result in repeated infections, as long as susceptible tissues are present. Perithecia form in late fall on old leaves, and asci mature in the spring to release ascospores without forcible discharge. The pycnidia formed on canes are larger and thicker-walled than those on leaves and do not form spores until they have overwintered.

Control

Management practices that increase air circulation within the planting and the row (such as proper spacing of plants, thinning to provide proper cane density, and maintaining narrow rows) are beneficial in controlling leaf spot. These practices promote faster drying of foliage and canes after rain and lessen the time available for infection to occur. Pruning out old fruiting canes and removing dead and damaged ones after harvest is helpful in reducing the amount of overwintering inoculum.

The fungicide spray programs used for controlling fruit rots (primarily Botrytis gray mold) are generally sufficient to control leaf spot in most situations. Additional sprays of benomyl, captan, and fixed-copper fungicides have been used when the disease is severe.

Selected References

Darrow, G. M. 1935. Susceptibility of raspberry species and varieties to leaf spot (*Mycosphaerella rubi*) at Beltsville, Maryland. Phytopathology 25:961-962.

Demaree, J. B. and Wilcox, M. S. 1943. The fungus causing the so-called "Septoria leaf-spot disease" of raspberry. Phytopathology 33:986-1003.

Roark, E. W. 1921. The Septoria leaf spot of *Rubus*. Phytopathology 11:328-333.

Wolf, F. A. 1935. The perfect stage of *Cercospora rubi*. Mycologia 27:347-356.

Zeller, S. M. 1937. Two Septoria leaf-spot diseases of *Rubus* in the United States. Phytopathology 27:1000-1005.

(Prepared by M. A. Ellis)

Septoria Leaf Spot of Blackberry

Septoria leaf spot affects leaves and canes of erect and trailing blackberries. It is generally accepted that the fungus causing this disease does not infect raspberry. It is a common and often severe disease in the southeastern United States and the Pacific Northwest, where it is often called cane and leaf spot. When severe, it can cause premature defoliation in late summer and early fall, resulting in reduced plant vigor and increased susceptibility to winter injury.

Symptoms

On leaves, the fungus produces typical frogeye lesions, with a whitish center and a brown or purple margin (Plate 25). The lesions are more circular in outline and larger and usually develop later in the season than those caused by anthracnose, with which they are often confused. Leaf spots caused by *Septoria* are about 3–4 mm in diameter. Small black pycnidia can be seen in the central portions of the lesions. Lesions on canes and petioles are similar to those on leaves but more elongate in outline.

Causal Organism

This disease is caused by the fungus *S. rubi* Westend. A teleomorph has been described as *Mycosphaerella rubi* Roark, but Demaree and Wilcox questioned the accuracy of this nomenclature. They also reported several synonyms for *S. rubi*, including names of species and varieties in *Septoria*, *Ascochyta*, and *Sphaerella*. *S. rubi* can be separated from *S. darrowii* Zeller (anamorph of the causal fungus for raspberry leaf spot) by conidia morphology. Although raspberry leaf spot is commonly referred to as Septoria leaf spot, it is caused by *Sphaerulina rubi* Demaree & M. S. Wilcox. These two fungi are reported not to cross-infect raspberry and blackberry.

Septoria rubi produces pycnidia that are brown to black, flattened, and epiphyllous and have wide ostioles. Conidia are long, slender (40–55 × 1.5 μm), and obscurely two- to three-pluriseptate. According to Roark (whose work could not subsequently be repeated), perithecia (60–80 μm in transverse diameter) are mainly hypophyllous, gregarious, erumpent, and globose. They lack paraphyses and have short papilliform ostioles. The perithecial walls are black, two or three cells thick, and pseudoparenchymatous. Asci (45 × 8–10 μm) are subclavate to cylindrical, very short-pedicellate, eight-spored, and irregularly biseriate. Ascospores (20–25 × 3.5–4.3 μm) are hyaline, slenderly fusiform, and straight to slightly curved. They are one-septate, with a very slight constriction at the septum.

Disease Cycle and Epidemiology

According to Roark, the fungus overwinters as mycelium and immature pycnidia in dead leaves and stems and, to a lesser extent, as perithecia. The fungus commonly moves into new fields as pycnidia on canes of nursery stock. Conidia are moved to infection courts by splashing or wind-driven rain. Ascospores are formed from May to July in Wisconsin. Secondary infections continue throughout the growing season and are related to periods of rainfall. Although the teleomorph is reported to occur in North Carolina, others have not found it in the southeastern United States and conclude that the fungus overwinters in the pycnidial stage. Roark did not successfully complete a cycle of controlled inoculations from ascospore to conidia and back to ascospore to unequivocally establish the relationship between *M. rubi* and *Septoria rubi*. The identity of the ascomycetous and related imperfect fungi causing the leaf spots of *Rubus* requires more study to reconcile the conflicting reports of the fungi involved.

Control

Management practices that increase air circulation within the planting and the row (such as proper spacing of plants, thinning to provide proper cane density, and maintaining narrow rows) are beneficial in controlling leaf spot. These practices promote faster drying of foliage and canes after rain and lessen the time available for infection to occur. Pruning out old fruiting canes and removing dead and damaged canes after harvest is helpful in reducing the amount of overwintering inoculum.

The fungicide spray programs used for anthracnose and fruit rots (primarily Botrytis gray mold) are beneficial in controlling Septoria leaf spot in the southern United States. In the South, a delayed-dormant spray of lime sulfur followed by three applications of ferbam or captan is the recommended schedule for leaf spot control on trailing blackberries. In the Pacific Northwest, a similar recommended three-spray schedule afforded some control on trailing blackberries but was not entirely satisfactory in years of severe infection. Bordeaux mixture in September and a winter-dormant application of lime sulfur were added to the spray schedule in Oregon in years of severe infection. The dormant sprays were found to be ineffective in California.

Selected References

Demaree, J. B., and Wilcox, M. S. 1943. The fungus causing the so-called "Septoria leaf-spot disease" of raspberry. Phytopathology 33:986-1003.

Diener, V. L., Eden, W. G., and Carlton, C. C. 1955. Leaf spot and strawberry weevil on trailing blackberries. Ala. Agric. Exp. Stn. Leafl. 46. 4 pp.

Roark, E. W. 1921. The Septoria leaf spot of *Rubus*. Phytopathology 11:328-333.

Wilhelm, S., Thomas, H. E., and Koch, E. C. 1951. Diseases of the Loganberry. Calif. Agric. 5:11-14.

Wolf, F. A. 1935. The perfect stage of *Cercospora rubi*. Mycologia 27:347-356.

(Prepared by M. A. Ellis)

Sydowiella and Gnomonia Cane Cankers

Sydowiella depressula (P. Karst.) Barr (syn. *Gnomonia depressula* P. Karst.) has been found in Scotland on canes of red raspberry growing in temporarily waterlogged soils in winter or in low-lying areas susceptible to frost injury. However, inoculation experiments in the field have failed to establish the pathogenicity of this fungus on red raspberry canes. In a major outbreak in 1979, a silver color developed on the lower portion (10–30 cm from the soil level) of overwintered canes, and few lateral shoots were produced on the upper third. The silvered epidermis was slightly rough in texture from numerous black beaks of immersed perithecia. When the bark was scraped away, water-soaked black or brown vascular lesions, which girdled the vascular cylinder or killed only the phloem, were revealed. Many canes with damaged phloem produced ripe fruit, but others wilted before harvest.

The immersed perithecia of *S. depressula* (250–480 × 150–400 μm) have extremely long beaks (150–450 [occasionally up to 1,200] × 60–180 μm) and contain unitunicate asci (60–90 × 11–16.4 μm), which are ellipsoid to oblong, with a broad, nonamyloid apical annulus and eight two-celled ascospores (26–31 × 7.9–10.5 μm). The smooth-walled, hyaline ascospores are ellipsoid to fusoid, with a median septum constricting the guttulate spore. A single gelatinous apical appendage at the end of the spore is visible when it first matures but may disappear with time.

S. depressula has also been recorded on *Rubus parviflorus* Nutt. in British Columbia and on dead stems of *Rubus* spp. in Europe and North America. It is possible that climatic or edaphic factors may cause the death of canes or predispose them to colonization by this organism.

Gnomonia rubi (Rehm) Winter in Rabenh. also occurs on *Rubus* cane fruits and has been shown by inoculation experiments to be pathogenic to stems of the blackberry cultivars Theodor Reimers (in Germany), Thornfree, Dirksen Thornless, and Hull Thornless (in the United States). It can also infect roses, particularly during cold storage before planting.

G. rubi can be distinguished from *S. depressula* by its smaller ascospores and the maturation of only four ascospores per ascus. The globose perithecia (165–280 × 125–250 μm) have shorter beaks (130–470 × 80–100 μm) than *S. depressula*. The unitunicate asci (33.5–44 × 7.5–13 μm), with a nonamyloid apical ring, contain ellipsoid ascospores (10–15.5 × 2.5–3.5 μm) with a median septum and setose apical appendages 1.5–6 μm long. An eight-spored form, named *G. rostellata* (Fr.:Fr.) Bref., has also been recorded on *Rubus* in North America and the United Kingdom.

G. rubi was more pathogenic on floricanes than on primocanes of thornless blackberry cultivars inoculated in the field in Ohio. There is a possibility that canes are predisposed to infection by winter injury, or that floricanes are inherently more susceptible.

No control measures for *G. rubi* or *S. depressula* have been described, and further studies are required to elucidate the possible role of predisposing factors, such as winter injury or waterlogging, in infection.

Selected References

Barr, M. E. 1978. The Diaporthales in North America with emphasis on *Gnomonia* and its segregates. Pages 57-58 in: Lehrte, W. Germany: J. Cramer.

Ellis, M. A., Kuter, G. A., and Wilson, L. L. 1984. Fungi that cause cane cankers on thornless blackberry in Ohio. Plant Dis. 68:812-815.

Schneider, R., Paetzholdt, M., and Willer, K.-H. 1969. *Gnomonia rubi* als Krankheitserreger an Kuhlhausrosen und Brombeeren. Nachrichtenbl. Dtsch. Pflanzenschutzdienstes (Stuttgart) 21:17-21.

Williamson, B. 1980. *Sydowiella depressula* on red raspberry. Trans. Br. Mycol. Soc. 74:647-649.

(Prepared by B. Williamson)

Nectria Canker of Raspberry

The causal organism of Nectria canker of raspberry is *Nectria mammoidea* W. Phillips & Plowr. var. *rubi* (Osterw.) Weese (anamorph *Cylindrocarpon ianthothele* var. *ianthothele* Wollenw). This fungus was described by Osterwalder in 1911 as a new species, *N. rubi*, causing a disease of raspberry roots in Switzerland. Weese (1912) reduced the taxon to varietal status as *N. mammoidea* var. *rubi*. Subsequently, disease outbreaks were reported in Ireland in 1916 and in Scotland and England during 1925–1926. Inoculation trials have failed to reproduce disease symptoms, and it is likely that the fungus is a secondary pathogen of plants subjected to stress such as wind damage followed by waterlogging.

In 1988, outbreaks of the disease occurred on red raspberries in Scotland after the previous year's severe summer gales, which compacted soil around primocanes and damaged them at the point of attachment to the crown. Many overwintered canes died before the buds opened, or lateral shoots wilted later in spring. Primocane production was reduced or absent in some plants. Dark brown or purplish vascular lesions at the crown caused the death of canes and poor primocane growth. In subsequent years these plantations returned to full health.

Clusters of perithecia form on the root collar or sometimes on the root below the soil. The perithecia are red when young and darken to purple or almost black when mature. Perithecia (300–500 μm in diameter) are smooth-walled and ovoid to globose, with a slightly protruding, darkened apical ostiole, through which ascospores extrude in white or buff tendrils. The perithecial wall lacks a pseudoparenchymatous structure and is composed of a complex, interwining network of thickened hyphae. Asci (90–130 × 6–9 μm) are cylindrical, with a rounded apex containing an indistinct ring. Asci contain eight ellipsoid ascospores (12–17 × 5–7 μm) with a single central septum. Ascospores are initially smooth-walled and hyaline, but at maturity may become faintly spinulose and pale brown.

On potato-sucrose agar, colonies reach 10–20 mm in diameter in 10 days at 20°C and usually have a strong purple pigmentation. The fungus also forms abundant *Cylindrocarpon* macroconidia in slimy purple sporodochia. Conidiogenous cells (10–25 × 2–3.5 μm) are cylindrical, with a collarette and thickened ring at their apex, and are terminally borne on irregularly branching cells. Conidia (36–58 × 4–5.5 μm) are hyaline, curved, fusoid to cylindrical, and three- to five-septate, with rounded apical and basal cells. They lack the oblique foot cell found in *Fusarium* macroconidia. *Nectria* perithecia may sometimes occur in agar culture.

Selected References

Alcock, N. L. 1925. A note on raspberry canker (*Nectria rubi* Osterwalder). Trans. Proc. Bot. Soc. Edinburgh 29:197-198.

Duncan, J. M. 1979. *Cylindrocarpon ianthothele* Wr. Page 84 in: Rep. Scott. Hortic. Res. Inst. 1978.

Natrass, R. M. 1927. Notes on *Nectria rubi* II. Trans. Br. Mycol. Soc. 12:23-27.

Osterwalder, A. 1911. Über eine neue auf kranken Himbeerwurzeln vorkommende *Nectria* und die dazu gehörige *Fusarium*-Generation. Ber. Dtsch. Bot. Ges. 29:611-622.

Pethybridge, G. H. 1927. Notes on *Nectria rubi* I. Trans. Br. Mycol. Soc. 12:20-23.

Ruokola, A. L. 1982. Fungus disease of raspberry (*Rubus idaeus* L.) in Finland. J. Sci. Agric. Soc. Finl. 54:99-111.

(Prepared by D. Brayford)

Silver Leaf

Silver leaf has occasionally been reported to affect *Rubus* spp. It was first noted on red raspberry in New Zealand in 1939, and other outbreaks on raspberry have been found in the United Kingdom and Norway, where the disease was associated with silvering of the foliage, poor growth, and death of canes. More recently, silver leaf was confirmed on blackberry in the United Kingdom.

The causal organism, *Chondrostereum purpureum* (Pers.:Fr.) Pouzar (syn. *Stereum purpureum* Pers.:Fr.), attacks a wide range of deciduous woody hosts, including apple, pear, cherry, peach, gooseberry, and black currant, but is most destructive on plum. On these hosts infection always starts with a wound in the bark of the stem or branch. It may also grow as a saprophyte on dead stumps, trunks, or branches of several trees and shrubs. On dead wood the fungus produces resupinate basidiocarps (2–8 cm), which are white and hairy (tomentose) on the upper surface. The spore-bearing underside (hymenium) of the basidiocarp is smooth and dark purple or brown to violaceous and fades with age. In vertical slices, the white upper layer is separated from the lower layers by a conspicuous dark line visible to the naked eye. The basidia (50 × 50 μm) have four sterigmata and are packed together into a dense hymenium. The spores (5–8 × 2.5–3 μm) are allantoid to subcylindrical, smooth-walled, and hyaline. One of the characteristic features is the presence of thin-walled cystidia (60–80 × 6–8 μm) that protrude from the surface of the hymenium up to 25–50 μm. The cystidia are smooth, but they may be covered with crystalline deposits.

Specific control measures for the disease on *Rubus* spp. have not been reported. On apple and plum, the protection of wounds with a suitable paint is helpful in limiting infection. Infected prunings on which the fungus could sporulate should be removed from the planting and destroyed. Trees such as birches, willows, and poplars are very susceptible to silver leaf. In areas where the disease has been a problem, such trees in close proximity to *Rubus* plantings should be monitored for the presence of the disease.

Selected References

Brien, R. M., and Atkinson, J. D. 1942. The occurrence of *Stereum purpureum* on the raspberry in New Zealand. N.Z. J. Sci. Technol. Sect. A 23:343-348.

Eriksson, J., and Ryvarden, L. 1973. The Corticiaceae in northern Europe. Pages 237-237 in: *Aleurodiscus* to *Confertobasidium*. Vol. 2. Fungiflora, Oslo.

Wilcox, H., and Gladders, P. 1980. Silver leaf of blackberry. Plant Pathol. 29:98.

(Prepared by T. Locke)

Fruit and Flower Diseases Caused by Fungi

Botrytis Fruit Rot (Gray Mold) and Blossom Blight

Gray mold, or Botrytis fruit rot, is the most common and one of the most serious diseases of *Rubus* spp. worldwide. The fungus rots fruit in the field before harvest (preharvest rot), particularly after persistent rain during blossom, but mainly causes rapid and devastating losses of picked fruits after harvest (postharvest rot). Occasionally it blights blossoms, but it rarely rots green or immature berries. Both blossom blight and gray mold are favored by a cool, moist environment. When such conditions persist during harvest, heavy losses in fruit yield and quality can be sustained. *Rubus* fruits, especially red raspberry, are extremely delicate and perishable and must be stored at high humidity after picking to reduce water loss. They have a shelf life of only a few days, even if refrigerated, and gray mold further shortens it. Gray mold is the single most important factor limiting the sale of fresh fruit to distant markets. The fungus also attacks senescent leaves, giving rise to cane infections known as cane Botrytis (see Cane and Foliar Diseases Caused by Fungi).

Symptoms

The first signs of the pathogen in late winter are superficial sclerotia on infected canes (Plate 12). These germinate in spring and become covered with masses of conidia. Closed flowers are generally not attacked, but once they are opened, flowers may be blighted, particularly after a late spring frost. Under moist conditions, the fungus may then profusely sporulate on dead flowers. As berries ripen on the plant, a few infected drupelets sometimes develop a watery rot and appear tan or golden tan. This water-soaked stage is transitory, and the drupelets soon become covered by a grayish brown, dusty mass of hyphae and conidia; eventually the entire fruit may be affected (Plate 26). This dusty gray covering gives the disease its name *gray mold*. Rot may develop on any part of the fruit but is common on drupelets near the pedicel (stalk end). Any type of physical damage results in a higher incidence of gray mold, especially during rainy periods. Common sources of injury include harvesting procedures, wind, spine abrasion, and insects. If left unharvested, the infected berries become mummified and remain attached to the receptacle on the plant, even in mechanically harvested fields.

Normally, the receptacles (tori) of picked fruit dry up and shrivel on the plant after the berries are picked, but during wet weather they are readily colonized by the fungus, and abundant conidia are produced. Sclerotia may also form in colonized receptacles.

In the high relative humidity necessary for postharvest storage, mycelial growth on the surface of infected berries is cottony and whitish to gray. Little sporulation occurs if light is excluded, but under diffuse light the fungus eventually sporulates as the fruit shrivels. Infected berries begin to leak, and adjacent ones become colonized.

Causal Organism

The disease is caused by *Botryotinia fuckeliana* (de Bary) Whetzel, but the anamorph (*Botrytis cinerea* Pers.:Fr.) is the only state recorded on *Rubus* spp. In culture, *B. cinerea* grows rapidly on several media, but the growth is variable. On potato-dextrose or potato-sucrose agar, the culture is initially whitish but quickly becomes brownish gray. Hyphae are branched, septate, and hyaline. Exposure to daylight or UV radiation is necessary for sporulation. Conidiophores (often 2 mm long or more and 16–30 μm thick) arise directly from the mycelium, usually have a swollen basal cell, and are dark in color. They are straight and alternately branched, mostly in the apical region. Both conidiophores and their branches show determinate growth. The terminal cells of branches swell into clavate ampullae, and conidia are formed as blastospores on very short denticles on the surface of the ampullae. Conidia (8–14 × 6–9 μm) form synchronously and are ellipsoid or obovoid, with a smooth surface (Fig. 13). In mass, they appear grayish brown, but individually they are colorless to pale brown. They separate from the denticle by a transverse septum, which often leaves a short neck or frill surrounding the hilum. Conidia are multinucleate, having one to 10 nuclei (the average number is four).

Sclerotia are considered to be the most important long-term survival structure of the pathogen, but the mycelium inside dead plant tissue is also important. Conidia are relatively short-lived, and the role of microconidia in survival and pathogenesis is uncertain. Sclerotia are produced mainly in lesions on infected canes; most strains of *B. cinerea* produce them in culture. They are black, round to irregular, and up to 5 mm across. Their shape in culture differs markedly from those formed on infected canes.

Disease Cycle

B. cinerea is a facultative parasite with numerous hosts, including weed species, and it survives as a saprophyte on plant debris. Nevertheless, the contribution of inoculum from sources outside raspberry fields is probably minimal. Conidia arising from sclerotia on canes and mycelia in dead leaves and mummified berries are the main sources of primary inoculum (Fig. 14 and Plate 27). They are dislodged by the twisting action of conidiophores and dispersed mainly by wind, but also by splashing water (rain and overhead irrigation). Unopened flowers are rarely colonized, because sepals are not readily infected. As soon as flowers open they become susceptible. Conidia landing on the sticky stigma germinate in the stigmatic fluid, and hyphae grow intercellularly throughout the transmitting tissues of the style and enter the ovary walls of many carpels within 7 days. Pollen does not need to be present for conidia to germinate on the stigma. The anther and filament of stamens also become colonized. The receptacle

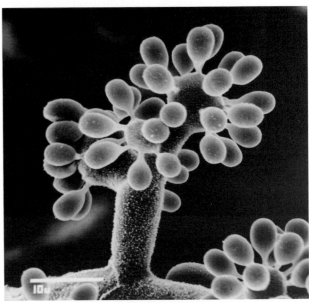

Fig. 13. Scanning electron micrograph of conidiophore and conidia of *Botrytis cinerea*. (Courtesy D. Lorenz)

in the center of the fruit remains sterile until exposed at harvest. Petals also become infected, but because most of them are shed soon after pollination, it is unlikely that they are important in gray mold development.

These initial infections of flower parts are considered latent. Except in the rare cases of flower blight, the fungus lies dormant inside immature fruit tissues and dead attached styles and stamens until the fruit is nearly ripe (or after harvest), at which time symptoms develop rapidly at high humidity. *B. cinerea* can also sporulate on infected stamens and stigmas close to harvest; hence secondary inoculum may be produced on developing berries in the absence of macroscopic symptoms. As the fruit ripens, the latent mycelium resumes activity, and the typical fruit rot symptoms develop. The direct infection of mature fruit is considered to be of little importance.

Epidemiology

Gray mold fruit rot is favored by cool, wet conditions, especially rainfall or overhead irrigation prior to and during the harvest period. The more mature or overripe the fruit becomes, the greater the chance that berries will rot and become covered with a felty mass of conidia. The concentration of airborne conidia depends initially on the amount of sporulating material available. Spore concentration is generally low during the spring, even though sclerotia germinate then, and it remains low until a considerable proportion of the fruit ripens. As it ripens, spore concentrations rise and remain elevated even after the ripe fruit is harvested. This has been attributed to sporulation on fleshy exposed receptacles, but the presence of unpicked overripe fruit is the most important factor in many plantings.

Wind speed, temperature (or radiation), relative humidity, and duration of surface water all profoundly affect the performance of *B. cinerea* as a pathogen because they affect the production, dispersal, deposition, and germination of conidia. Jarvis, in early studies of raspberries in Scotland, showed that conidia are produced during still, cool periods at night and dispersed by wind in the morning when the temperature rises and the relative humidity falls. This pattern is reversed in the evening. Several recent epidemiological studies of *B. cinerea* in tomatoes, grapes, snap beans, and strawberries confirmed this diurnal rhythm of spore dispersal and showed that most conidia were produced at 21°C and 94% RH with a wind speed of 0.6 m/sec. A poor correlation was found in Scotland between the number of spores caught in a raspberry plantation and the incidence of postharvest gray mold. This may be due to the infection of stigmas from relatively few conidia, even in the absence of rain or dew, and the establishment of latent infections in styles.

Control

Control can be achieved only by breaking the disease cycle by means of chemical and cultural methods and the use of plant resistance.

Protectant fungicide sprays are applied at 7- to 14-day intervals from early bloom up to (and sometimes into) harvest. Captan (in North America) and dichlofluanid (in Europe) have been extensively used for control. More recently MBC-generating compounds (benomyl, carbendazim, and thiophanate-methyl) and dicarboximides (vinclozolin and iprodione) have been used. *B. cinerea* has developed resistance to most of these fungicides in many areas. To prolong their usefulness, it is generally recommended that different classes of fungicide be applied alternately or as tank mixes. Exclusive use of dicarboximide fungicides may lead to the buildup of other postharvest fruit rots. *Cladosporium* sp. and *Rhizopus* sp. are insensitive to dicarboximides and may replace gray mold as the dominant postharvest rots.

Cultural practices that create an open plant canopy improve air circulation, increase light penetration, and speed the drying of plant surfaces after rain or irrigation. Such practices also minimize the premature senescence of lower leaves due to shading and reduce infection. Rigorous pruning to maintain a narrow row in a hill (stool) system, avoidance of excess nitrogen fertilizer, and efficient weed control similarly help to maintain an environment less conducive to gray mold. Control of primocane vigor by removal of the first flush of primocanes, mechanically or chemically (cane burning), allows fruit to ripen in an open canopy.

Training systems that create an open canopy also reduce the risk of gray mold and other diseases and improve the penetration of fungicide sprays. Worldwide there are numerous training methods, each of which has its own merits in different environments and for various economic reasons.

Biennial cropping (alternate-year bearing) should reduce the incidence of fruit rot, because primocanes are absent in the fruiting year, but infection of primocanes in the vegetative season may be severe unless they are sprayed. Primocane-fruiting types eliminate overwintering canes and hence should lower the primary inoculum. These types, however, are prone to preharvest rot, because their fruit matures later in the summer, when rainfall is greater and temperatures are lower.

Red raspberry cultivars with improved resistance to preharvest and postharvest gray mold have been developed, but none show exceptional resistance. Hence, the use of a resistant cultivar does not eliminate the need for other control measures. Important characteristics associated with resistance include the upright habit of lateral shoots and fruit that is spaced evenly along the laterals, separates easily from the receptacle when ripe, and has a tough skin and firm texture.

Postharvest fruit rot. Postharvest gray mold is important mainly for fresh-market berries, but fruit for processing may also rot rapidly before freezing, juicing, or canning, causing a reduction in quality. Fungicides applied between early bloom and harvest reduce the incidence and delay the onset of postharvest fruit rot. A series of interrelated practices used in the handling of fresh *Rubus* fruits have been developed in the Pacific Northwest to reduce the incidence of rotting. The main objectives are to avoid or minimize injury to fruit and to maintain an environment that slows its ripening and the growth of *B. cinerea*.

In the Pacific Northwest, pickers are specially trained and paid a bonus to harvest berries at the correct stage of ripeness. The fruit is picked at the red ripe stage of maturity, when it is very firm, and extra force must be used to pull it from the receptacle; at this stage it continues to ripen but will not increase in size. The berries are picked directly into the containers that will be sold to the customer, and to avoid crushing

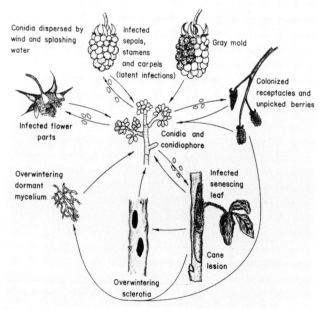

Fig. 14. Disease cycle of *Botrytis cinerea* (gray mold) on red raspberry fruit.

only shallow containers are used. Because the fruit ripens quickly, it is picked daily for fresh market when the weather is warm, or every second day when it is cooler. Harvesting in early morning, when the berries are firm and cool, and rapid transport to the cooler are essential.

The fruit is cooled to near 0°C as soon as possible after harvest to retain its firmness during refrigerated storage. When it is covered with cellophane, temperature control is critical to prevent condensation. Overall, this practice tends to lead to an increase in the incidence of gray mold, but overwrapping is requested by many retail outlets. It is vitally important to maintain a temperature of about 2°C throughout the storage, distribution, and sale of *Rubus* fruits. It is equally important to maintain high humidity in storage to prevent desiccation.

Selected References

Coley-Smith, J. R., Verhoeff, K., and Jarvis, W. R., eds. 1980. The Biology of *Botrytis*. Academic Press, New York. 318 pp.

Dashwood, E. P., and Fox, R. A. 1988. Infections of flowers and fruit of red raspberry by *Botrytis cinerea*. Plant Pathol. 37:423-430.

Ellis, M. B., and Waller, J. M. 1974. *Sclerotinia fuckeliana* (conidial state: *Botrytis cinerea*). Descriptions of Pathogenic Fungi and Bacteria, No. 431. Commonwealth Mycological Institute, Kew, Surrey, England.

Jarvis, W. R. 1962. The dispersal of spores of *Botrytis cinerea* Fr. in a raspberry plantation. Trans. Br. Mycol. Soc. 45:549-559.

Jarvis, W. R. 1977. *Botryotinia* and *Botrytis* Species: Taxonomy, Physiology and Pathogenicity. A Guide to the Literature. Monogr. 15. Agriculture Canada, Ottawa. 195 pp.

Johnson, K. B., and Powelson, M. L. 1983. Analysis of spore dispersal gradients of *Botrytis cinerea* and gray mold disease gradients in snap beans. Phytopathology 73:741-746.

McNicol, R. J., Williamson, B., and Dolan, A. 1985. Infections of red raspberry styles and carpels by *Botrytis cinerea* and its possible role in post-harvest grey mould. Ann. Appl. Biol. 106:49-53.

McNicol, R. J., Williamson, B., and Dolan, A. 1990. Effects of inoculation, wounding and temperature on post-harvest gray mold (*Botrytis cinerea*) of red raspberry. J. Hortic. Sci. 65:157-165.

Williamson, B., and McNicol, R. J. 1986. Pathways of infection of flower and fruits of red raspberry by *Botrytis cinerea*. Acta Hortic. 183:137-141.

Williamson, B., McNicol, R. J., and Dolan, A. 1987. The effect of inoculating flowers and developing fruits with *Botrytis cinerea* on post-harvest grey mould of red raspberry. Ann. Appl. Biol. 111:285-294.

(Prepared by P. R. Bristow)

Postharvest Soft Rot (Leak Disease)

The shelf life of harvested fruits of *Rubus* spp. during cold storage is limited primarily by the onset of fungal spoilage. Apart from losses from gray mold, caused by *Botrytis cinerea* Pers.:Fr., the principal postharvest disorder of *Rubus* fruits is soft rot, or leak disease, caused by *Rhizopus* and *Mucor* spp.

Symptoms

Rhizopus and *Mucor* spp. infect only damaged or mature fruit directly; they do not infect flowers. Symptoms are rarely observed in the field except on late-season, overripe berries. During storage, initial symptoms of fruit infection may include water-soaking, but the most obvious sign is the development of mycelial growth on the fruit surface. Affected berries eventually deteriorate and release juices. The infection may spread very rapidly from fruit to fruit or between storage containers, and as it spreads, the characteristic leaking of juice from storage containers occurs, and hence the common name *leak disease*.

These fungi produce webs of coarse white to gray hyphae, which spread rapidly over the surface of the fruit. The weblike hyphae become covered with erect, slender white hyphae that support conspicuous black sporangia. This condition is often referred to as *pin mold*.

Causal Organisms

The principal causal organism of soft rot, or leak disease, in North America is *R. stolonifer* (Ehrenb.:Fr.) Vuill. (syn. *R. nigricans* Ehrenb.), although its distribution is worldwide. Mycelia spread very rapidly between fruit by means of stolons, which are implanted by rhizoids into the substrate at each node. Internodes may be 1–3 cm long. Sporangiophores are 1–2 mm long, abundantly produced, and usually united in groups of three to five. Each sporangiophore supports a terminal hemispherical sporangium, 85–200 μm in diameter. Sporangia contain sporangiospores (10–20 × 7.5–8 μm) that are dark brown to black, unequal, angular, and irregularly round to oval. *R. stolonifer* is heterothallic, and compatible strains produce black, verrucose, and round or oval zygospores, 160–220 μm in diameter.

In Europe the principal cause of soft rot is *M. piriformis* E. Fisch., which has dense, silver-gray mycelium. Its sporangiophores are 800–200 μm long and up to 70 μm thick and may be produced singly or in small groups. Black, spherical sporangia (80–200 μm in diameter) contain hyaline, ovoid sporangiospores (6–7 × 9–12 μm). *M. piriformis* is also heterothallic, and compatible strains produce large, black, verrucose, spherical zygospores.

Two other fungi, *R. sexualis* (Smith) Callen and *M. hiemalis* Wehmer, can also cause soft rot but are less frequently isolated. Mycelial growth of *R. sexualis* is generally sparse and close to the surface of infected fruit. The fungus is homothallic, and the surface of infected fruit rapidly becomes covered with black, spiny, globose zygospores, 80–100 μm in diameter. Sporangia are pale, globose, and borne on trailing sporangiophores (up to 1500 μm in length) and contain pale, gray, and angular to ovoid sporangiospores (5–12 μm in diameter). *M. hiemalis* can commonly be isolated from the surface of healthy fruit but is less frequently isolated from diseased fruit. Its mycelial growth is dense, yellowish gray, and composed of fine, slender hyphae. The spherical sporangia (50–80 μm in diameter) are borne on erect sporangiophores and contain hyaline, oval sporangiospores (5–10 × 2.5–5 μm).

Disease Cycle

Both *Mucor* and *Rhizopus* spp. overwinter as saprophytic mycelium and as dormant, asexual spores in infested leaf debris in the *Rubus* planting. Neither species sporulates on leaf debris. Initial infection is, therefore, generally slow and requires contact between ripe or damaged fruit and dormant spores or mycelium in debris. Once the fruit becomes infected, inoculum builds up rapidly. This partially explains why infection is usually not observed in the field, and why postharvest infection is generally more prevalent on fruit harvested late in the season. Late-season raspberries have been reported to be more susceptible to infection than early-season, but the rate of spoilage of stored blackberries and loganberries appears not to differ with the time of season.

Disease incidence varies markedly from season to season, suggesting that it is greatly influenced by climatic conditions. The transfer of propagules by rain-splash is particularly important for *M. piriformis*, and rainstorms during the fruiting season have a marked effect on the development of infections and subsequent spoilage caused by *Mucor* spp. In contrast, under high humidity, *R. stolonifer* is predominantly mycelial in habit, whereas *R. sexualis* produces abundant zygospores (the function of which is unclear, since they have not been observed to germinate). Both *Rhizopus* spp. produce abundant asexual sporangia when the humidity drops below 80%. *Rhizopus* spp. generally cause serious spoilage problems during warm, dry seasons, when the inoculum is at its greatest and propagules are readily spread by air currents.

Once established on ripe fruit in the plantation, inoculum can build up extremely rapidly, because of mycelial growth and abundant spore production. During storage, infection spreads rapidly as a result of contact between healthy and infected fruit and with the dispersal of fungal propagules.

Infections on raspberries often develop within the cavity left by the removal of the receptacle (torus) during harvest. Juice exudes from deteriorating drupelets into the cavity, creating an ideal humidity and nutrient supply for the growth of these fungi. This is generally not a problem on blackberries or loganberries, from which the receptacle is not removed at harvest.

Infection by all of these fungi results in rapid maceration of the fruit (caused by the production of pectolytic enzymes) and in soft rot or leak. The juices released from such tissues may also carry fungal propagules, thus providing an additional means of pathogen dispersal.

Infection by *Mucor* spp. can occur at temperatures as low as 0°C. Growth and infection by *Rhizopus* spp. ceases at temperatures below 5°C. The spores of *R. sexualis* rapidly lose viability on fruit held at 0°C and consequently cannot cause infections when the fruit is returned to higher temperatures.

Control

There are no chemical control methods for *Rhizopus* or *Mucor* spp. on *Rubus* fruits. Control must be achieved through cultural methods that reduce the incidence of fruit infection in the field and the rate of infection and spread during postharvest storage.

Debris is the most important source of primary inoculum. Consequently, it should be removed as thoroughly as possible from the planting prior to the fruiting season. Fruiting on low lateral shoots, which are close to the soil, should be avoided to prevent the rain-splash of inoculum from soil or plant debris. Any practice that improves air circulation through the fruiting canes should be encouraged. This aids surface drying of the fruit and hence reduces infection, especially by *Mucor* spp.

Firm, ripe fruit should be picked as regularly as possible and quickly cooled to as near to 0°C as possible to reduce the rate of infection by *Mucor* spp. and completely inhibit the growth of *Rhizopus* spp. The harvesting of overripe berries should be avoided; careful handling should minimize bruising, which can be a particular problem with mechanical harvesting. The fruit should be packed in shallow containers preferably made of material that absorbs any leaked juice.

Modified-atmosphere storage (with reduced oxygen and increased carbon dioxide levels) in conjunction with cold storage has had variable success, depending on the season as well as the harvest date. Fumigation with acetaldehyde has been reported to control the infection of raspberries by *R. stolonifer* without affecting fruit quality, but this has not been pursued commercially.

Clonal variation in the susceptibility of raspberry fruit to postharvest infection by *Rhizopus* spp. is associated with tissue resistance and fruit firmness. Relatively resistant genotypes, such as the cultivar Glen Prosen, may be useful parents for selection of more resistant cultivars. Similarly, raspberry clones derived from *Rubus crataegifolius* Bunge have been reported to produce fruit with a longer shelf life, thereby providing another potential source of resistance.

Selected References

Cohen, E., and Dennis, C. 1975. Effect of fungicides on the mycelial growth of soft fruit spoilage fungi. Ann. App. Biol. 80:237-242.

Daubeny, H. A., and Pepin, H. S. 1974. Variations among red raspberry cultivars and selections in susceptibility to the fruit rot causal organisms *Botrytis cinerea* and *Rhizopus* spp. Can. J. Plant Sci. 54:511-516.

Daubeny, H. A., Pepin, H. S., and Barritt, B. H. 1980. Postharvest Rhizopus fruit rot resistance in red raspberry. HortScience 15:35-37.

Davis, R. P., and Dennis, C. 1977. The fungal flora of loganberries in relation to storage and spoilage. Ann. Appl. Biol. 85:301-304.

Dennis, C. 1975. Effect of pre-harvest fungicides on the spoilage of soft fruit after harvest. Ann. Appl. Biol. 81:227-234.

Dennis, C., and Cohen, E. 1976. The effect of temperature on strains of soft fruit spoilage fungi. Ann. Appl. Biol. 82:51-56.

Dennis, C., and Mountford, J. 1975. The fungal flora of soft fruits in relation to storage and spoilage. Ann. Appl. Biol. 79:141-147.

Freeman, J. A., and Pepin, H. S. 1976. Control of pre- and postharvest rot of raspberries by field sprays. Acta Hortic. 60:73-79.

Knight, V. H. 1980. Responses of red raspberry cultivars and selections to *Botrytis cinerea* and other fruit-rotting fungi. J. Hortic. Sci. 55:363-369.

Lunn, J. A. 1977. *Rhizopus stolonifer*. Descriptions of Pathogenic Fungi and Bacteria, No. 524. Commonwealth Mycological Institute, Kew, Surrey, England.

(Prepared by R. P. Davis)

Minor Fruit Rots

Several fungi that have been associated with the spoilage of *Rubus* fruits during postharvest storage include *Alternaria* spp., *Cladosporium* spp., *Penicillium* spp., and *Colletotrichum gloeosporioides* (Penz.) Penz. & Sacc. in Penz. They have a widespread distribution, and many are saprophytes associated with fruit plantings. They can commonly be isolated from the surface of apparently healthy berries and are generally associated with late-season, overripe, or damaged fruit.

Alternaria rot. This disease occurs in mature raspberry fruit, usually after harvest. The infected berries are covered with dark gray mycelium, on which chains of dark, muriform conidia are borne. *A. humicola* Oudem. was one of two molds most commonly found on black raspberries harvested in Michigan.

Cladosporium rot. The two *Cladosporium* spp. principally associated with infection of *Rubus* fruits are *C. herbarum* (Pers.:Fr.) Link and *C. cladosporioides* (Fresen.) G. A. De Vries. Both produce olive green, velvety mycelial growth over the surface of affected fruit. Conidiophores are geniculate and up to 250 μm long and produce conidia both terminally and from nodes. Dark green conidia are produced by both fungi in long, often branched chains. Conidia of *C. herbarum* are 5–23 × 3–8 μm, ellipsoidal or oblong, distinctly verrucose, and nonseptate or one-septate. Conidia of *C. cladosporioides* are 4–25 × 2–6 μm, ellipsoidal, smooth or minutely verrucose, and mostly nonseptate.

Both *Cladosporium* spp. have a worldwide distribution and can easily be isolated from air, soil, and dead woody or herbaceous materials. They are abundant in *Rubus* plantings and can generally be isolated from the surface of apparently healthy berries. Their incidence on the surface of fruit is generally uniform throughout the season, although it tends to be higher after rainstorms.

Infected fruit becomes covered with a velvety green growth. Mycelial growth is generally restricted to the fruit surface, with little or no damage to the tissues, but the appearance of fungal growth renders the berries unfit for sale. Visible growth is usually restricted to overripe or damaged fruit and often associated with secondary growth on lesions caused by *Botrytis cinerea* Pers.:Fr. Growth of the fungus occurs optimally at 20–25°C and may continue at normal fruit storage temperatures.

Penicillium rot (blue mold). Several species of *Penicillium* have been implicated in the rot of *Rubus* fruits. These species are widespread and commonly found in plantings in soil, in plant debris, and on the surface of apparently healthy fruit. The occurrence of these fungi on the fruit surface also appears to increase after rainstorms. During storage *Penicillium* spp. appear on the surface of the fruit as a powdery fungal growth, which is initially white but soon turns blue-green (thus the name *blue mold*). Often the fungi are associated with bruised, damaged, or overripe fruit and develop in the cavity of raspberries. Infected tissue softens, and juice may leak from the berries.

Colletotrichum rot. The causal fungus of this disease, *C. gloeosporioides*, has a worldwide distribution and an extremely wide host range and has been associated with ripe fruit rots in a number of important crops. It is relatively common on the surface of ripe *Rubus* fruits, and its incidence is often greatest after rainstorms. Although the fungus is commonly isolated from the surface of ripe, healthy fruit, Colletotrichum rot is relatively rare. Infection is generally apparent only on stored fruit and can occur on the surface as well as in the cavity of raspberries. The lesion is characterized by a slightly sunken, water-soaked area, which often develops a slimy fungal growth in the center.

Control

There are no chemical control methods for the minor fruit rots of *Rubus*, but dichlofluanid, used for the control of Botrytis gray mold, has been somewhat effective with *Cladosporium* spp. Control is generally enhanced by good sanitation or hygiene within the planting, very regular and careful harvesting, and rapid cooling and storage of fruit at temperatures as close to 0°C as possible. Infected fruit should be removed before marketing.

(Prepared by R. P. Davis)

Stamen Blight

Stamen blight has been recorded throughout the Pacific Northwest (especially west of the Cascade Mountains) and in Canada, Denmark, Germany, Great Britain, Ireland, and Spain. It has occurred in cultivated blackberry, boysenberry, cascadeberry, evergreen blackberry, loganberry, red raspberry, and youngberry, as well as in the wild on nine species of *Rubus*, including red raspberry.

A high incidence of diseased flowers threatened to decrease the yield of cultivated dewberries in Oregon. Diseased fruit of boysenberry and cascadeberry in British Columbia were reported to be small and of little value. Although the loss in yield of cultivated red raspberry in Scotland was not significant, malformed fruit caused difficulties during harvest.

Symptoms

Diseased flowers differ from healthy ones by their white, powdery appearance. In some hosts, infected stamens lie flat against the petals and sepals, making the flowers appear larger and more conspicuous than healthy ones, in which the stamens stand upright around the stigmas (Plate 28). The white "powder" consists of masses of fungus spores originally produced on anthers and later scattered over stigmas, petals, and sepals. When sepals are removed from unopened flower buds, diseased anthers are already brown and their surfaces covered with white coils of spores. Generally, the flowers in an inflorescence are either all diseased or all healthy, and both types occur on the same cane. Intermediate stages, in which only some flowers in an inflorescence are diseased, are less conspicuous.

In wild blackberry, a phyllody of sepals, an increase in the number of petals, and a type of witches'-broom have been observed on diseased inflorescences. Witches'-broom has been associated with the disease in wild dewberry and wild trailing blackberry, but not in boysenberry, evergreen blackberry, youngberry, or red raspberry.

Berries that develop from diseased flowers vary from small, dry, and seedy to large and apparently normal (Plate 29). The most frequent symptom occurs between these two extremes—berries in which only some drupelets have developed normally. Many berries from diseased flowers are more firmly attached than normal to the receptacle, which makes harvesting difficult. When handpicking is employed, it results in crushed, wet fruit unsuitable for fresh markets.

Causal Organism

Hapalosphaeria deformans (Syd.) Syd. produces light brown, rounded or conical pycnidia approximately 50–80 µm in diameter. Pycnidia are produced mainly in the walls of the anthers on infected flowers, but they may be found at a later stage in the style, petals, sepals, and fruit. Smooth, hyaline, nonseptate, and spherical or ovoid spores 3–5 µm in diameter are extruded from the pycnidia as white, mucilaginous coils. The fungus mycelium is intercellular, septate, and 2–4 µm in diameter and has been observed to extend into the receptacle and pedicels of blackberry, but not raspberry.

Disease Cycle and Epidemiology

The spores produced on diseased flowers of fruiting canes may be dispersed by rain, mechanical agitation, bees, or wind throughout the flowering period of the host. Primocanes may become infected when spores are deposited in the leaf axils containing the buds that will develop into inflorescences the following year. When water is present in the leaf axils, spores are trapped between the developing scales of the axillary bud. Sparse, nonpenetrating mycelium is formed within the bud during winter and early spring, and penetration of the anthers and some of the carpels occurs in late spring. Mycelial growth within the anthers is rapid, and after 2 weeks pycnidia have exuded spores onto the surface of anthers within the unopened flower bud. Infected anthers never dehisce, and consequently pollen is not released. Infection has a direct, adverse effect on fruit size and drupelet development and also affects pollen production.

Epidemics may occur when the percentage of diseased flowers is high and conditions are suitable for spore dispersal and infection of axillary buds. Bud infection is most important, since the severity of disease in one year does not necessarily determine that of the following year. In general, disease severity increases as a plantation ages, but wide fluctuations may occur from year to year.

Although diseased flowers occur at all nodes in the cultivated raspberry, the greatest number appear at the eighth and ninth nodes from ground level. As a result, the maximum amount of inoculum is likely to be available for dispersal when this part of the cane produces flowers. The first rainfall disperses the greatest number of spores from opened flowers. More than 60% of the spores are deposited in the vicinity of the diseased plant, but some are deposited at at least 3.7 m away. Wind and bees may disseminate spores, but in fewer numbers than those from rain-splash, and the probability of the former methods of dispersal resulting in infections of axillary buds is much less.

No matter how concentrated the inoculum is, the infection of axillary buds in raspberry is dependent on the retention of water around them for a sufficient period to enable the spores to become trapped between the scales of the developing buds. These conditions are most likely when there is vigorous growth, primocanes are numerous, and air movement within the row is reduced.

Cross-infections can occur between wild and cultivated raspberries and between wild blackberry and cultivated raspberry.

Control

Where the disease is prevalent in red raspberry, it is essential to establish new plantings with disease-free plants from a commercial nursery. Care must be taken to remove wild *Rubus* spp. from around the planting, because they may serve as a source of inoculum for introducing the disease. Despite these precautions, inoculum sources some distance from the planting may initiate infections when spores are disseminated by wind and rain.

When stamen blight is established within the planting, fungicide application may be required. The disease can be significantly decreased by annual applications of fungicides commonly used for control of Botrytis fruit rot. Some of the nozzles on the spray boom may need adjusting to direct the fungicide downward, onto the axillary buds of the primocanes. The spray volume should be sufficient to allow for runoff, determined by the vigor and density of the primocanes. Sprays applied during the second and third weeks after the start of flowering have resulted in a significant decrease in the number of diseased inflorescences the following year. In one study, a single application of lime sulfur in August was beneficial in controlling the disease on youngberry during the following season.

Selected References

Blyth, W. 1949. Studies on *Hapalosphaeria deformans* Sydow. Ph.D. thesis. University of Edinburgh. 77 pp.

Chamberlain, G. C., and Putnam, W. L. 1955. Diseases and insect pests of the raspberry. Can. Dep. Agric. Publ. 880 (rev.). 34 pp.

Dickens, J. S. W. 1967. Epidemiology of stamen blight of raspberry. Ann. Appl. Biol. 60:343-351.

Dickens, J. S. W. 1968. The pathological histology of stamen blight of raspberry. Trans. Br. Mycol. Soc. 51:519-524.

Montgomerie, I. G. 1971. Control of raspberry stamen blight. Ann. Appl. Biol. 67:321-329.

Zeller, S. M., and Braun, A. J. 1943. Stamen blight of blackberries. Phytopathology 33:136-143.

(Prepared by I. G. Montgomerie)

Rust Diseases

Orange Rust

Orange rust is most important in northeastern North America. It has also been recorded on *Rubus* spp. elsewhere in the United States and Canada and in Europe, Asia, and Australia. It is caused by a systemic rust confined to *Rubus* spp. (autoecious) but exists in two forms with separate names—a long-cycled (*demicyclic*) form, which predominates on black raspberries, and a short-cycled (*endocyclic*) form, which predominates on blackberries and dewberries. The short-cycled form is a common and serious pathogen of erect and trailing blackberries throughout the United States. Red raspberry is the only *Rubus* sp. reported to be immune to the disease.

Orange rust is not known to kill plants, but it causes considerable reductions in fruit production and vegetative growth. Infected plants rarely recover, which results in total loss to the producer. Commercial plantings in the United States with 10-25% of the plants infected have reported serious decreases in fruit and plant production. Because rust-infected plants are of no value, the losses from this disease are of considerable importance.

Symptoms

Symptoms on infected plants can be seen as soon as new growth appears in the spring. Young shoots are usually spindly and clustered. The unfolding leaves are usually stunted or misshapen and pale green to yellowish (Plate 30). Spermagonia develop on the upper surface of the leaves even before the leaves unfold. They originate from fungal intercellular mycelium in the subepidermal leaf tissue and at maturity appear as black specks surrounded by chlorotic leaf tissue. Approximately 3 weeks later, the lower leaf surface becomes covered with blisterlike aecia (Plate 31). These are waxy at first but soon turn powdery and bright orange as they rupture and shed aeciospores. On black raspberries, the rusted leaves start to wither and drop in late spring to early summer.

This disease is marked by a proliferation of shoots from infected roots, instead of single, strong shoots arising from buds on healthy roots. The infected shoots lack spines and are weak and spindly. In June, many primocanes and floricanes on infected plants appear healthy, but the entire plant nevertheless becomes systemically infected. A plant remains infected throughout its life, and its floricanes produce no blossoms.

In mid- to late summer, plants infected with the long-cycled rust develop brownish black pustules (telia) on the lower surface of leaves, generally from ground level up to about 60 cm.

Causal Organisms

Two organisms cause orange rust on *Rubus* cane fruits: *Arthuriomyces peckianus* (E. Howe) Cummins & Y. Hiratsuka (the long-cycled form) and *Gymnoconia nitens* (Schwein.) F. Kern & H. W. Thurston (the short-cycled form, as described by Cummins and Hiratsuka). Previously both forms of the rust were combined under the single species *G. nitens* (syn. *G. peckiana* (E. Howe) Trott.). Morphological differences between these organisms have been reported. Both produce spermagonia on the upper leaf surface, although a strain of the short-cycled form that lacks them has been found. The two forms can be distinguished by germinating the aeciospores on water agar. The aeciospores (16–24 × 19–30 μm) are globoid, with finely verrucose, hyaline walls (Fig. 15). *G. nitens* produces reddish orange, waxy aecia containing smaller and more irregular aeciospores, which produce a basidium. These aeciospores, therefore, function as teliospores by producing four-celled basidia, from which four basidiospores are produced. In contrast, *A. peckianus* has golden orange aecia, with aeciospores that germinate to produce only long germ tubes. Neither species produces urediniospores; *A. peckianus*, however, produces true teliospores. The telia (0.2–1.0 mm in diameter) appear on the underside of older leaves as small, dark brown to black spots. The teliospores (18–27 × 32–45 μm) are two-celled and smooth.

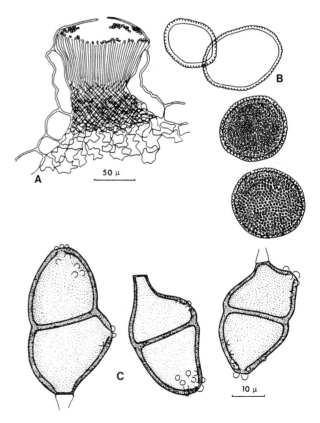

Fig. 15. Pycnia (**A**), aeciospores (**B**), and teliospores (**C**) of *Gymnoconia nitens* (orange rust). (Reprinted, by permission, from Laundon and Rainbow, 1969)

Disease Cycle and Epidemiology

In late May to early June, windborne and perhaps rain-splashed aeciospores from *A. peckianus* are deposited on leaves of *Rubus* plants, where they infect only localized areas of individual mature leaves (Fig. 16). When environmental conditions favorable for infection occur, the spores germinate, producing germ tubes with appressoria. An infection peg soon develops from the appressorium and penetrates the cuticle and epidermis. The mycelium then colonizes the intercellular spaces of the spongy mesophyll, and haustoria penetrate the mesophyll cells. About 21–40 days after infection, small, brownish black telia develop on the underside of infected leaflets. The teliospores borne in these telia germinate to produce a basidium, which in turn produces basidiospores. In blackberries these spores infect buds on cane tips as they root; they also infect buds or new shoots being formed at the crowns of healthy plants in the summer. Both fungi become systemic, grow down into the crown at the base of the infected shoot, and enter newly formed roots. As a result, a few canes from the crown will show rust the following year. These fungi overwinter as systemic, perennial mycelium within the host. *G. nitens* possibly overwinters as teliospores.

Orange rust is generally favored by low temperatures and high humidity. Temperatures ranging from 6 to 22°C favor penetration and further development of the fungus, but higher temperatures at the time of inoculation decrease the percentage of germination. At 25°C, aeciospores germinate very slowly, and disease development is greatly retarded. Spore germination and plant penetration have not been observed at 30°C. Light is not a determining factor in the germination of spores. Aeciospores require long periods of leaf wetness before they can germinate, penetrate, and infect plants.

Fully developed leaves are more susceptible to aeciospore infection than young ones. Thus, leaf maturation should be considered if inoculations are made for pathogenicity studies or for screening seedlings or clones for resistance to orange rust.

Control

Present control recommendations rely almost entirely on cultural practices. It is important to establish new plantings from rooted cuttings that are free of rust and to remove infected wild *Rubus* spp. nearby. Subsequent roguing to remove and dispose of infected plants from within the planting is also necessary during early spring, before orange rust pustules on infected leaves mature and release spores. Any management practice that improves air circulation within the planting (such as thinning out canes within the row, pruning out floricanes immediately after harvest, and effective weed control) aid in disease control by reducing the duration of wetness.

There are no black raspberry cultivars resistant to orange rust. Purple raspberries have been reported to be susceptible it, but red raspberries are immune. The blackberry cultivars Eldorado, Raven, and Ebony King are reported to show resistance.

Fungicidal sprays used for the control of other *Rubus* diseases may be beneficial in reducing the number of new infections of orange rust but generally do not provide adequate disease control. Recent research with ergosterol biosynthesis–inhibiting fungicides indicates that they may be effective for controlling the disease on *Rubus* spp.

Selected References

Cummins, G. B., and Hiratsuka, Y. 1983. Illustrated Genera of Rust Fungi. American Phytopathological Society, St. Paul, MN. 152 pp.

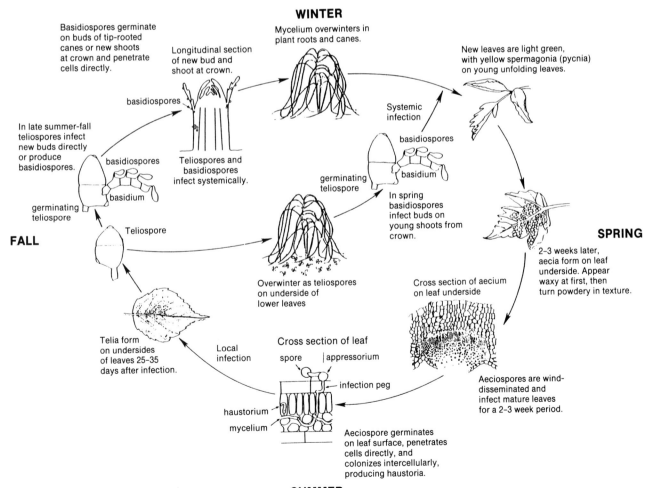

Fig. 16. Disease cycle of orange rust, caused by *Gymnoconia nitens*.

Dodge, B. O. 1923. Systemic infections of *Rubus* with the orange rusts. J. Agric. Res. 25:209-242.

Kleiner, W. C. 1988. Evaluation of fungicides for the management of orange rust on black raspberry. M.S. thesis. Pennsylvania State University, University Park. 96 pp.

Laundon, G. F., and Rainbow, A. F. 1969. *Gymnoconia nitens*. Descriptions of Pathogenic Fungi and Bacteria, No. 201. Commonwealth Mycological Institute, Kew, Surrey, England.

Pady, S. M. 1935. Aeciospore infection in *Gymnoconia interstitialis* by penetration of the cuticle. Phytopathology 25:453-474.

(Prepared by W. C. Kleiner and J. W. Travis)

Cane and Leaf Rust

On susceptible blackberry cultivars, cane and leaf rust can be of economic importance in the southeastern United States and the Pacific Northwest. The susceptibility of blackberry cultivars and hybrids varies greatly. Both red and black raspberries are recorded hosts, but occurrence of the disease on raspberries is rare. Although cane and leaf rust is not a systemic rust, it is often mistaken for the systemic orange rusts caused by *Arthuriomyces peckianus* (E. Howe) Cummins & Y. Hiratsuka and *Gymnoconia nitens* (Schwein.) F. Kern & H. W. Thurston. It is important to be able to distinguish between the two rust diseases in order to implement the proper control recommendations. Severe infection can lead to premature defoliation, which reduces plant vigor and makes plants more susceptible to winter injury. The infection of fruit can also occur, though rarely.

Symptoms

Cane and leaf rust is first seen in late spring on infected floricanes, which develop large lemon yellow uredinia that split the bark (Plate 32). Small yellow uredinia form during early summer on undersurfaces of leaves on floricanes (Plate 33). In years of severe infection this can cause premature defoliation. Uredinia rarely occur on the fruit. Buff-colored telia develop among the uredinia on leaves in early fall. Spermagonia and aecia also form on lower leaves of primocanes in October and November.

Causal Organism

Cane and leaf rust is caused by the fungus *Kuehneola uredinis* (Link) Arth. Spermagonia are epiphyllous on reddish spots, large, prominent, and pustular. Orange-yellow aecia surround the spermagonia, often in confluent rings. Aeciospores (18–19 × 19–23 μm) are globoid or obovoid, and the walls are colorless, closely verrucose, and 2–2.5 μm thick, with obscure pores. Uredinia are hypophyllous, scattered, powdery, and lemon yellow when fresh. Old parasitized uredinia may be white. The urediniospores (16–19 × 21–27 μm) are obovoid (Fig. 17). Their walls are nearly colorless, finely and closely verrucose-echinulate, and 1.5–2 μm thick and have three to four indistinct equatorial pores. Telia are hypophyllous, scattered among uredinia on old leaves, and pale buff. Teliospores (18–24 × 85–110 μm) are cylindrical, five- to 13-celled, irregularly flattened or coronate above, and narrow below, with each cell extending into a pore-bearing tip next to the cell above; their walls are colorless, 1.5–2 μm thick on the sides, thicker above, and smooth or slightly roughened at the apex. Pedicels are colorless and very short to apparently lacking.

Disease Cycle and Epidemiology

The fungus probably overwinters on canes as mycelium or latent uredinia, but the details of infection and overwintering are unclear. Urediniospores from lesions on the floricanes infect leaves on the same or other floricanes during the growing season. Disease development is favored by wet conditions. Telia develop on leaves of floricanes in the fall, and basidiospores from germinating teliospores infect adjacent leaves of primocanes where spermagonia and aecia are formed.

Control

To reduce the inoculum, old diseased canes should be pruned and removed after fruiting. In Oregon, a three-spray program is recommended for cane and leaf rust control. Lime sulfur is applied in winter, and two fixed copper sprays are applied at the green tip stage and again just before blooming. Spraying and sanitation have not provided satisfactory control of cane and leaf rust in Washington State.

Care must be taken to distinguish between blackberry plants infected with the systemic orange rust fungus, which have pustules of orange waxy aeciospores on leaves only, and those infected with the cane and leaf rust fungus, which have yellow pustules of powdery urediniospores on floricanes as well as on leaves. Plants infected with orange rust must be removed (including the roots) from the planting and destroyed. This drastic control procedure is not required for the nonsystemic cane and leaf rust.

Selected References

Arthur, J. C. 1934. Manual of the Rusts of the United States and Canada. Purdue Research Foundation, Lafayette, IN. 438 pp.

Fischer, G. W., and Johnson, F. 1950. Cane and leaf rust, *Kuehneola uredinis* (Link) Arth., of blackberries in western Washington. Phytopathology 40:199-204.

Laundon, G. F., and Rainbow, A. F. 1969. *Kuehneola uredinis*. Descriptions of Pathogenic Fungi and Bacteria, No. 202. Commonwealth Mycological Institute, Kew, Surrey, England.

(Prepared by M. A. Ellis)

Yellow Rust

Yellow rust is usually considered to be a relatively minor rust disease of red raspberry. It has a worldwide distribution wherever raspberries are grown. Severe infections that cause yield losses are exceptional and confined to highly susceptible cultivars grown in humid areas. The disease can cause premature defoliation if the infection is early and severe, and the consequent loss of winterhardiness can reduce yields substantially. In the Pacific Northwest, young canes can be attacked—hence such other common names for the disease as *cane rust* and *western yellow rust*.

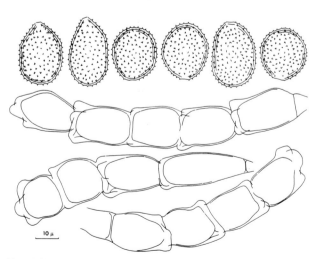

Fig. 17. Urediniospores (top) and teliospores (bottom) of *Kuehneola uredinis* (cane and leaf rust). (Reprinted, by permission, from Laundon and Rainbow, 1969)

Symptoms

The first symptom of yellow rust in spring is bright orange-yellow aecia on the upper surfaces of leaves on young canes and lateral shoots close to the ground (Plates 34 and 35). Occasionally aecia can be seen on petioles, peduncles, and sepals before and after flower buds burst. In June or July orange to pale yellow uredinia appear on the underside of leaves and sometimes on pedicels, sepals, and drupelets (Plate 36). Uredinia have been recorded on young canes in Oregon but are uncommon on canes in Britain. The yellow uredinia darken as black teliospores form from mid-July through late autumn. The underside of leaves on susceptible cultivars may become densely covered by the black telia. In Oregon, it has been reported that overwintering cane lesions become deep and cankered, and that affected fruiting canes may break or dry up in the following summer. This rust is not systemic. The production of orange-yellow aecia only on the upper surfaces of leaves is a simple character that may be used to distinguish it from late leaf rust, caused by *Pucciniastrum americanum* (Farl.) Arth.

Causal Organism

Raspberry yellow rust is caused by the fungus *Phragmidium rubi-idaei* (DC.) P. Karst. (syn. *P. imitans* Arth.), which is an autoecious macrocyclic rust confined largely to *Rubus idaeus* L. subsp. *vulgatus* Arrhen. and *R. idaeus* subsp. *strigosus* (Michx.) Maxim. The spermagonia appear as small (less than 1 mm in diameter), bright orange, slightly elevated spots on the upper surface of leaves in the spring. Most spermagonia occur on leaves of young canes and on lateral shoots, generally low in the canopy. As many as 100 spermagonia may be produced per leaf. Aecia develop as conspicuous orange rings (0.5–2.0 mm in diameter), with spermagonia in the center and inconspicuous peripheral paraphyses. Often two concentric rings of aecia form, with the outer aecial ring developing several days after the inner. The aecial walls are evenly thin. Aeciospores (15–27 μm in diameter) are globose to ellipsoid; their walls are hyaline and sparsely echinulate, with obscure pores. The yellow uredinia (0.1–0.4 mm in diameter) are hypophyllous and uncurved and have numerous peripheral paraphyses. The urediniospores (17–23 × 15–19 μm) are broadly obovoid, with hyaline walls (1.25–1.5 μm thick) that are strongly echinulate and have obscure pores (Fig. 18). Paraphyses (50–90 × 15–28 μm) are clavate and uncurved, with the wall thickened to 2–3 μm at the apex. The six- to nine-celled teliospores are cylindrical and not constricted at the septa. They are rounded at the apex and usually have a small umbo or are tapered to a hyaline apiculus up to 20 μm long. The teliospores (28–32 × 79–124 μm excluding the apiculus) have sienna to dark chestnut brown, thick walls (3.5–5.5 μm) with a coarsely warted surfaced and three pores in each cell. The hyaline pedicels that support the teliospores are 100–170 μm long and 8–11 μm wide at the neck, swelling to a width of 17–22 μm at the base. The pedicels swell in water, become sticky, and aid in the attachment of the teliospores to canes.

Disease Cycle and Epidemiology

The fungus overwinters only as teliospores. The most abundant source of teliospores in spring is the surface of the bark on floricanes, particularly in natural splits near the base, where up to 400 spores per square centimeter have been recorded. Teliospores require several months of winter weather to overcome dormancy. They germinate above 6.3°C (optimum 14.6–18.4°C and maximum 25°C) to produce promycelia bearing four basidiospores (Fig. 19). Germination occurs most efficiently in darkness. In the United Kingdom spermagonia appear in a single flush, suggesting that teliospore germination and the production of basidiospores occur during a brief period of favorable conditions. Aecia are usually present for a period of 4–5 weeks in spring and early summer. Aeciospores initiate new infections on leaves by penetration of stomata. Urediniospores germinate only in water above 10.8°C (optimum 18.4–20.9°C and maximum 25°C) and penetrate the leaf through stomata. The disease can progress rapidly, as successive flushes of urediniospores are produced during extended periods of cool, wet weather. Only the youngest green tissues are susceptible, and free water must be present on these tissues before urediniospores can germinate and penetrate the plant. In Oregon it has been reported that deeply penetrating lesions on young canes predispose them to infection by the cane blight fungus, *Leptosphaeria coniothyrium* (Fuckel) Sacc. Two phys-

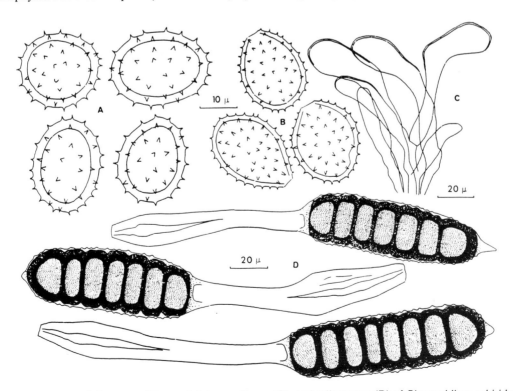

Fig. 18. Aeciospores (**A**), urediniospores (**B**), uredinial paraphyses (**C**), and teliospores (**D**) of *Phragmidium rubi-idaei* (yellow rust). (Reprinted, by permission, from Laundon and Rainbow, 1969)

iologic races of *P. rubi-idaei* have been reported in Washington State, and tests indicate that variation in pathogenicity is present in populations of the fungus in the United Kingdom.

Control

The use of resistant red raspberry cultivars in areas where the disease is often severe is an important method of control. Major gene resistance to *P. rubi-idaei* is present in the North American cultivars Latham, Chief, and Boyne, and strong incomplete resistance of the slow-rusting type is present in the cultivar Meeker. The British cultivar Malling Jewel is rarely affected, because it shows strong resistance that functions against the uredinial and telial stages.

Any practice that provides increased air circulation in the row or between plants (such as "stooling up" or cane thinning) should aid disease control by allowing tissues to dry more rapidly.

The complete removal of the first flush of primocanes, either by desiccant herbicides or by hand (cane burning or cane vigor control), is a method widely used to reduce excessive cane vigor in some cultivars, and it effectively controls the disease in the United Kingdom. The removal of primocanes destroys the fungus in the initial stages of its life cycle before any symptoms appear, and with no other source of inoculum the disease is effectively controlled. Primocanes produced later in the spring escape the disease, provided the entire planting is treated simultaneously.

Biennial cropping (alternate-year bearing) can also control the disease, because the complete removal of floricanes and old cane stubs removes most of the teliospore inoculum. An application of lime sulfur as a late-dormant spray has been reported to be effective. Benodanil has provided effective control in the United Kingdom but can only be applied postharvest.

Selected References

Anthony, V. M., Shattock, R. C., and Williamson, B. 1985. Life-history of *Phragmidium rubi-idaei* on red raspberry in the United Kingdom. Plant Pathol. 34:510-520.

Anthony, V. M., Shattock, R. C., and Williamson, B. 1985. Interaction of red raspberry cultivars with isolates of *Phragmidium rubi-idaei*. Plant Pathol. 34:521-527.

Anthony, V. M., Williamson, B., Jennings, D. L., and Shattock, R. C. 1986. Inheritance of resistance to yellow rust (*Phragmidium rubi-idaei*) in red raspberry. Ann. Appl. Biol. 109:365-374.

Anthony, V. M., Williamson, B., and Shattock, R. C. 1987. The effect of cane management techniques on raspberry yellow rust (*Phragmidium rubi-idaei*). Ann. Appl. Biol. 110:263-273.

Johnson, F. 1946. Physiologic races of yellow rust of raspberries in western Washington. Phytopathology 36:383-384.

Laundon, G. F., and Rainbow, A. F. 1969. *Phragmidium rubi-idaei*. Descriptions of Pathogenic Fungi and Bacteria, No. 207. Commonwealth Mycological Institute, Kew, Surrey, England.

Zeller, S. M. 1927. The yellow rust of raspberry caused by *Phragmidium imitans*. J. Agric. Res. 34:857-863.

Zeller, S. M., and Lund, W. T. 1934. Yellow rust of *Rubus*. Phytopathology 24:257-265.

(Prepared by B. Williamson)

Late Leaf Rust

Late leaf rust (also called autumn rust, late raspberry rust, late yellow rust, and American spruce-raspberry rust) affects mainly cultivated red and purple raspberries and some wild red raspberries. It occurs in California, British Columbia, and the northern parts of central and eastern North America. In the past the disease was considered to be of minor importance, but recently has been serious outbreaks have occurred, particularly in California and the Atlantic provinces. In severe cases leaf infections cause premature defoliation, which reduces plant vigor and increases the susceptibility of canes to winter injury. Fruit is also infected and may be rendered unfit for fresh-market sales by the presence of powdery yellow spore masses on the drupelets. In summer-bearing cultivars the fruit often escapes infection, and the disease probably goes unnoticed in many plantings, because it does not develop until after harvest. However, severe outbreaks at harvest occur occasionally (e.g., in Nova Scotia, where a planting of the summer-bearing red raspberry cultivar Festival had 70% of the fruit infected). The fruit of fall-bearing cultivars does not escape late-season disease development: in an outbreak in Ohio, 30% of the fruit from a planting of the cultivar Heritage was infected and unfit for sale.

Symptoms

The fungus that causes late leaf rust is not systemic. Therefore, late leaf rust differs considerably from orange rust of black raspberry and blackberry, caused by *Arthuriomyces peckianus* (E. Howe) Cummins & Y. Hiratsuka and *Gymnoconia nitens* (Schwein.) F. Kern & H. W. Thurston. On mature leaves, many small spots develop and turn yellow and eventually brown before the leaves die in the fall. Small uredinia filled with powdery yellow spores (not waxy ones, like orange rust spores) are formed on the underside of infected leaves (Plate 37). Badly infected leaves drop prematurely, and the canes of highly susceptible cultivars may be bare by September. Such canes are more susceptible to winter injury. Flower calyces, petioles, and fruit at all stages of development are also attacked. On fruit, uredinia develop on individual drupelets, producing yellow masses of urediniospores, which make the berries unsightly and unfit for fresh-market sales (Plate 38). Dodge reported the occurrence of late leaf rust lesions on raspberry canes in Maryland, but they have not been found on canes in Nova Scotia.

Causal Organism

Late leaf rust is caused by the fungus *Pucciniastrum americanum* (Farl.) Arth. This heteroecious, macrocyclic rust produces spermagonia and aecia on white spruce (*Picea glauca*

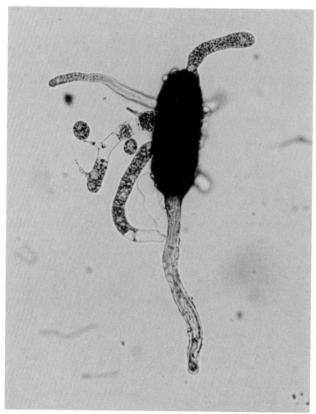

Fig. 19. Germinating teliospore of *Phragmidium rubi-idaei*, showing promycelia and basidiospores. (Courtesy B. Williamson; copyright Scottish Crop Research Institute. Used by permission)

(Moench) Voss) (Plate 39) and uredinia and telia on *Rubus* spp. White spruce is the most common host for the aecial stage; Engelmann spruce (*Picea engelmanii* Parry ex Engelm.) has also been reported to be susceptible. The main uredinial hosts are red raspberries (*R. idaeus* subsp. *melanolasius* Focke and *R. idaeus* subsp. *strigosus* Michx.) and purple raspberries (*R. × neglectus* Peck). Other species reported to be susceptible are *R. leucodermis* Douglas ex Torr. & A. Gray, *R. occidentalis* L., and *R. ursinus* Cham. & Schlechtend.

Spermagonia develop in groups on the needles of white spruce and are subcuticular, cone-shaped, and 70–100 μm in diameter. Aecia develop on the current year's needles and occasionally on cone scales. They are cylindrical, peridiate, and 0.2–0.5 mm in diameter. Aeciospores (13–22 μm in diameter) are ellipsoid to obovoid (Fig. 20). Their walls are hyaline and verrucose, except for an almost smooth area at each end of the spore. They have an outer layer of spines up to 1.5 μm thick and an inner layer 0.5–0.75 μm thick. Uredinia are hypophyllous, fructicolous, scattered, and often crowded over large areas. Occasionally they are also formed on calyces and petioles. Uredinia are pale yellowish and peridiate. The peridium is cone-shaped or cylindrical, 70–110 μm in diameter, and 70–110 μm high. It has four to six ostiolar cells, which are laterally free, constricted in the middle, 23–31 μm high and 13–15 μm broad, smooth below, and aculeate above, with large warts 3–4 μm high and 2.5–3 μm broad. Urediniospores (18–27 × 11–17 μm) are obovoid to ellipsoid, with bright yellow contents. Their walls are hyaline and echinulate, with spines that are 2.5–3 μm apart and 0.75 μm high by 1–1.5 μm thick. Telia are amphigenous, scattered, and indistinct. Teliospores (15–22 × 15–20 μm), embedded just beneath the epidermis, are globose to ellipsoid. They contain two or more cells separated by vertical septa; each cell is 6.5–8.5 μm wide, with smooth, brownish walls about 1 μm thick.

A closely related but physiologically distinct rust fungus, *Pucciniastrum arcticum* Tranzschel, also occurs on *Picea glauca* and northern *Rubus* spp., such as *R. arcticus* L. *P. americanum* and *P. arcticum* do not cross-infect their respective *Rubus* hosts. Although Dodge considered the morphological variability in *P. americanum* to be great enough to include *P. arcticum* types, he did not reduce the two species to synonymy. Laundon and Rainbow noted that the two species differ in that the ostiolar cells of the uredinia of *P. americanum* are constricted and laterally free.

Physiologic specialization in *P. americanum* is unknown.

Disease Cycle and Epidemiology

Nickerson and Mahar studied the disease cycle and epidemiology of late leaf rust in Nova Scotia, where white spruce is common. Aecia begin to mature on infected spruce needles in mid-June (Fig. 21). Aeciospores are released in mid-June to early July and are capable of infecting raspberry during this period. In summer-bearing red raspberry cultivars, such as Festival, the first uredinia usually appear on flower calyces and the leaves of fruiting laterals in early July. Urediniospores can cause new infections on raspberry throughout the growing season. By late July, uredinia are found on fruit and the lower leaves of floricanes and primocanes. The younger leaves of primocanes are usually the last to show symptoms and may escape infection, as the rust epidemic slows with the onset of cooler weather in early fall. Telia develop in infected leaves in the fall, and teliospores germinate during rainy periods from mid-May to early June the following year. Basidiospore release and the infection of spruce needles coincide with the period of rapid expansion of the spruce buds.

In Nova Scotia, severe rust outbreaks in plantings of summer-bearing red raspberries at harvest (late July to mid-August) are likely in years when conditions favor the development of aecia on nearby spruce earlier in the season. This suggests that aeciospores are an important source of primary inoculum in these plantings. However, minor rust outbreaks may occur on summer-bearing cultivars during or after harvest, even in years when conditions are unfavorable for aecial development, and in plantings far removed from spruce. Several investigators have noted that late leaf rust apparently does not need the alternate host in order to survive on raspberry, because the disease is found year after year in regions remote from spruce trees. It seems likely that the uredinial mycelium is capable of overwintering in raspberry canes and producing urediniospores that serve as a source of primary inoculum for new infections the following year. In Nova Scotia, however, no rust lesions have been found on raspberry canes. Further research is needed on the survival of the uredinial stage and its role in the disease cycle of late leaf rust.

Both aeciospores and urediniospores of *P. americanum* are disseminated by wind. Urediniospores may also be mechanically spread from infected to healthy fruit during harvest. They are capable of infecting leaves of the susceptible red raspberry cultivar Comet at temperatures as low as 8°C, but the disease is more severe at 18–26°C. They germinate, and the germ tubes penetrate stomata mainly on the lower leaf surface in approximately 6 hr at 20°C. Free water on the leaf surface is not required for infection, but high humidity is essential. Frequent rainfall and sprinkler irrigation favor disease development. Middle-aged leaves on actively growing plants are most susceptible to infection. Very young leaves, older ones that are beginning to senesce, and those on plants under environmental stress or entering dormancy are less susceptible. In highly susceptible cultivars, such as Comet, uredinia are produced in

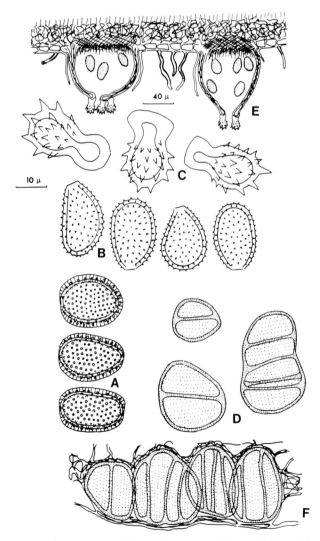

Fig. 20. Aeciospores (**A**), urediniospores (**B**), uredinial ostiolar cells (**C**), teliospores (**D**), and cross sections of uredinia (**E**) and telia (**F**) of *Pucciniastrum americanum* (late leaf rust). (Reprinted, by permission, from Laundon and Rainbow, 1969)

abundance and are mature 6–7 days after infection. In resistant cultivars, uredinia may be sparse or nonexistent and their maturation delayed. The fruit is susceptible to infection at all stages of development.

Control

In the past, late leaf rust was of minor importance and did not warrant special control measures. Recently, however, the disease has become much more prevalent and in certain locations has caused significant damage, particularly on the summer-bearing red raspberry cultivar Festival and the fall-bearing cultivar Heritage.

Any practice that increases air circulation in the planting (such as thinning canes, maintaining narrow rows, and controlling weeds) should aid in the control of late leaf rust by allowing susceptible tissues to dry more quickly. In areas such as the Atlantic provinces of Canada, where white spruce (the alternate host) plays a role in the disease cycle, the removal of leaves and other debris from infected raspberry plantings in the fall should help to break the cycle by reducing spruce infection the following spring. The eradication of white spruce near existing raspberry plantings is seldom practical except on a small scale, but growers should avoid establishing new plantings near stands of white spruce. In areas where the uredinial stage may survive the winter, the removal of old floricanes and infected primocanes during the winter should reduce the amount of overwintering inoculum.

The fungicides currently registered for use on raspberries are not very effective against the rust fungi. In Canada, applications of anilazine, timed to coincide with aeciospore release from white spruce, reduced the incidence of late leaf rust in experimental plantings of the cultivar Festival.

The prospects for control of late leaf rust through the use of resistant cultivars appear promising. In Canada, the summer-bearing cultivar Nova was highly resistant to *P. americanum* both in the field and in controlled-environment studies.

Selected References

Anderson, H. W. 1956. Diseases of Fruit Crops. McGraw-Hill, New York. pp. 326-327.

Arthur, J. C. 1934. Manual of the Rusts in the United States and Canada. Purdue Research Foundation, Lafayette, IN. p. 13.

Darker, G. D. 1929. Cultures of *Pucciniastrum americanum* (Farlow) Arthur and *P. arcticum* (Lagerheim) Tranzschel. J. Arnold Arbor. Harv. Univ. 10:156-167.

Dodge, B. O. 1923. Morphology and host relations of *Pucciniastrum americanum*. J. Agric. Res. 24:885-894.

Ellis, M. A., and Ellett, C. W. 1981. Late leaf rust on Heritage red raspberry in Ohio. Plant Dis. 65:924.

Laundon, G. F., and Rainbow, A. F. 1969. *Pucciniastrum americanum*. Descriptions of Pathogenic Fungi and Bacteria, No. 210. Commonwealth Mycological Institute, Kew, Surrey, England.

Luffman, M., and Buszard, D. 1988. Control of late yellow rust [*Pucciniastrum americanum* (Farl.) Arth.] of red raspberry. Can. J. Plant Sci. 68:1185-1189.

Nickerson, N. L., and Mahar, J. 1987. Late yellow rust of raspberries in Nova Scotia. Page 17 in: 1986 Annual Report, Agriculture Canada Research Station, Kentville, Nova Scotia.

Ziller, W. G. 1974. The tree rusts of western Canada. Environ. Can. Can. For. Serv. Publ. 1329. pp. 177-178.

(Prepared by N. L. Nickerson)

Blackberry Rust

Blackberry rust is a destructive disease of the European blackberry (*Rubus fruticosus* L.) and some cultivated blackberries common throughout Europe and the Middle East. It is also found in Chile, Australia, and New Zealand. The rust was deliberately introduced into Chile in 1972 and Australia in 1984 in attempts to control species of European blackberry that have become established as important weeds. Cultivated blackberries known to be susceptible include the cultivar Thornless Evergreen (*R. laciniatus* Willd.), which is highly

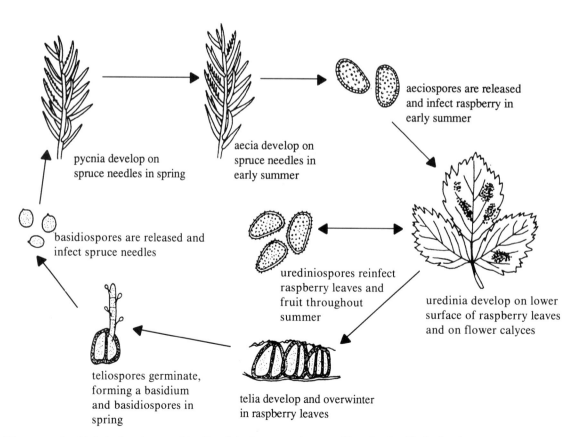

Fig. 21. Disease cycle of late leaf rust, caused by *Pucciniastrum americanum*. (Drawing by Cindy Gray)

susceptible, and Dirksen Thornless, Black Satin, Smoothstem, and Thornfree, which are moderately susceptible. Less susceptible cultivars include Comanche, Cheyenne, Lawton, Silvan, and Cherokee. Severe infections can result in premature defoliation, weakened cane growth, and reduced fruit production.

Symptoms

Blackberry rust can affect leaves, leaf veins, petioles, actively growing shoots, flowers, and fruit. The most common and obvious symptoms are on leaves. Shortly after infection, yellowish red blotches appear on the upper surface of infected leaves (Plate 40). Within about 10 days of infection they develop into purple or red circular spots (up to 4 mm in diameter), often with yellow or brown centers. On the lower leaf surface, directly beneath these spots, golden yellow, powdery uredinia up to 1 mm in diameter appear (Plate 41). Under conditions of severe disease development the edges may curl up, and the entire leaf may turn chlorotic and drop prematurely. On very susceptible hosts leaf spots are generally larger than on less susceptible ones. The golden pustules are replaced by black, powdery telia within a few weeks. On some highly resistant hosts the only symptom may be purple spotting on the upper leaf surface, with no spore production.

Causal Organism

Blackberry rust is caused by the fungus *Phragmidium violaceum* (C. F. Schultz) G. Wint., which is macrocyclic and autoecious. Uredinia and telia are the two most commonly observed spore-producing structures. Both are hypogenous and produced within the characteristic purple or red leaf spots. Uredinia are orange-yellow and surrounded by numerous hyaline, clavate, and incurved paraphyses. Urediniospores ($22-32 \times 19-24$ μm) are ellipsoid, with a hyaline wall echinulate with spines 3–5 μm apart and indistinct pores. Teliospores ($60-110 \times 31-38$ μm) are cylindrical and often constricted slightly at the septa. They have a rounded apical papilla and are one- to five-celled, dark brown, and covered with hyaline warts. They are borne on a pedicel (90–150 μm long), which is swollen at the base and up to 19 μm wide. Teliospores germinate to produce four-celled basidia, which bear up to four basidiospores. Spermagonia are epigenous, hemispherical, and crowded in the center of leaf spots; they produce oval spermatia. Aecia are hypogenous on leaf spots, orange-yellow, caeomatoid, and paraphysate. Aeciospores are similar in appearance and size to urediniospores.

Disease Cycle and Epidemiology

The blackberry rust fungus probably overwinters primarily as teliospores on old leaves persisting on the plant. However, at least one report indicates that it may overwinter as perennial mycelium on stems and produce urediniospores directly in the spring. Teliospores germinate in early spring to produce basidia, each of which bears up to four basidiospores. The basidiospores infect young leaves in the spring, and dark green spots develop on the upper leaf surface. Yellow spermagonia are produced in the center of these spots, and soon after, aecia are formed on the lower surface. Within a short period of time aeciospores follow, which cause additional infections on other leaves. Uredinia and urediniospores are produced on infected leaves about 10 days after aeciospore infections occur. Repeated infections resulting from urediniospores can develop on susceptible tissue throughout the growing season whenever favorable conditions exist. The pathogen is disseminated by windblown spores. In Chile, the disease spreads up to distances of 70 km within 1 year following its introduction. The conditions required for infection are not well defined, but temperatures of 18–20°C and free moisture on leaves for up to 18 hr allows abundant infection by urediniospores in inoculation studies. Young, rapidly growing tissue is most susceptible to infection, and resistance to the rust increases with leaf age.

Control

As with most foliar fungal pathogens, any practice that improves air movement within the plant canopy and shortens the drying time of the foliage will restrict the development of the disease. The control of wild blackberry plants growing among or near commercial plantings will reduce potential sources of inoculum and aid in disease control.

Detailed fungicide spray schedules have not been devised for this disease. However, where control measures are necessary, an early fungicide application as buds open followed by additional sprays during cane extension and flowering should prevent serious infection. Later, sprays for berry rot control may also prevent the spread of blackberry rust. Fungicides active against this disease include protectants (such as cupric hydroxide, mancozeb, and thiram) and systemic ergosterol biosynthesis–inhibiting fungicides (such as triforine, propiconazole, myclobutanil, and triadimefon). Some fungicides commonly used for gray mold control (including benomyl, iprodione, and vinclozolin) have little or no activity against the rust fungi.

Other rust fungi that may occur on European blackberry include *Kuehneola uredinis* (Link) Arth. and *P. bulbosum* (Strauss) Schlechtend. Though usually not economically important, they are often confused with blackberry rust, and their presence on wild blackberries may influence disease-management decisions because of the false threat of *P. violaceum* inoculum.

P. bulbosum has a host range and distribution similar to that of *P. violaceum* but can be distinguished from the latter by its smaller sizes of all the spore types. Aeciospores (17–21 μm) have coarsely verrucose walls, with broad, shallow warts up to 3 μm in diameter. Urediniospores ($19-24 \times 16-18$ μm) are echinulate, with spines 1.25–2 μm apart and approximately 0.5 μm high. Teliospores are also slightly smaller.

Selected References

Blackman, V. H. 1904. On the fertilization, alternation of generations and general cytology of the Uredineae. Ann. Bot. 18:323-372.

Bruzzese, E., and Hasan, S. 1987. Infection of blackberry cultivars by the European blackberry rust fungus, *Phragmidium violaceum*. J. Hortic. Sci. 62:475-479.

Loundon, G. F., and Rainbow, A. F. 1969. *Phragmidium bulbosum*. Descriptions of Pathogenic Fungi and Bacteria, No. 203. Commonwealth Mycological Institute, Kew, Surrey, England.

Loundon, G. F., and Rainbow, A. F. 1969. *Phragmidium violaceum*. Descriptions of Pathogenic Fungi and Bacteria, No. 209. Commonwealth Mycological Institute, Kew, Surrey, England.

Washington, W. S. 1987. Susceptibility of *Rubus* species and cultivars to blackberry leaf rust (*Phragmidium violaceum*) and its control by fungicides. J. Phytopathol. 118:265-275.

(Prepared by W. S. Washington)

Minor Rust Diseases

Arctic rusts. Two rusts infect cultivated arctic raspberry (*Rubus arcticus* L.) and other closely related subarctic berries and their hybrids in *R.* subg. *Cylactis* in Finland, Sweden, and Canada. Neither fungus appears to cause serious yield losses, but their incidence may be quite high toward the end of the growing season.

Pucciniastrum arcticum Tranzschel is a heteroecious macrocyclic rust that produces spermagonia and aecia on white spruce (*Picea glauca* (Moench) Voss) and uredinia and telia on *R. arcticus*, *R. saxatilis* L. (stoneberry), *R. pubescens* Raf., and *R. triflorus* Richardson. *P. arcticum* differs from *P. americanum* (Farl.) Arth., the cause of late leaf rust of raspberry, because the ostiolar cells of the uredinia in the latter are constricted and laterally free. The two rusts will not cross-infect their respective *Rubus* hosts.

Phragmidium arcticum Lagerh. is an autoecious rust on *R. arcticus* and related fruits. Spermagonia have not been reported and are probably lacking, and the uredinia are apparently scarce. The telia generally lack paraphyses, but in most respects the spore dimensions are similar to those of *P. rubi-idaei* (DC.) P. Karst. on red raspberry (see Yellow Rust). There is little information about the epidemiology of these two arctic rusts, and no control measures have been described.

Subtropical rust. *Hamaspora longissima* (Thüm.) Körn. is an autoecious rust affecting several *Rubus* spp., including *R. fruticosus* L. agg. and *R. lasiocarpus* Sm. in eastern South Africa, where it is widespread and common on wild *Rubus*. It also occurs in other African countries, such as Kenya, and in temperate hilly regions of India, China, and Taiwan. It has no economic importance in these areas but may present a risk to blackberries in other countries if introduced.

The teliospores are fusiform, hyaline, and septate, have long pedicels, and germinate without dormancy. They resemble teliospores produced in *Phragmidium* spp., except that they are hyaline rather than dark brown.

Selected References

Anonymous. 1977. *Hamaspora longissima* (Thüm.) Körn. Eur. Mediterr. Plant Prot. Organ. Data Sheets on Quarantine Organisms. EPPO List A1. 3 pp.
Cummins, G. B., and Hiratsuka, Y. 1983. Illustrated Genera of Rust Fungi. American Phytopathological Society, St. Paul, MN. p. 117.
Ryynanen, A. 1973. *Rubus arcticus* L. and its cultivation. Ann. Agric. Fenn. 12:1-76.
Savile, D. B. O. 1975. *Phragmidium arcticum*. Fungi Canadenses, No. 79. Natl. Mycol. Herb. Agriculture Canada. 2 pp.
Sydow, H., and Sydow, P. 1915. Monographia Uredinearum. Vol. 3 (3). p. 449.

(Prepared by B. Williamson)

Root and Crown Diseases Caused by Fungi

Phytophthora Root Rot

The first report of Phytophthora root rot on *Rubus* was in 1937, when the disease was described on red raspberry plants in Scotland. It was next reported in northwestern North America on loganberry and red raspberry in 1958 and 1965, respectively, and has since been considered a serious problem on the latter in that region. However, the disease was not recognized as a widespread problem until the mid-1980s, when serious outbreaks were documented on red raspberry in central and eastern North America, the United Kingdom, continental Europe, and Australia. Phytophthora root rot is now regarded as a major cause of declining red raspberry plantings in many commercial production areas worldwide. The relatively recent surge in importance of this disease appears to result from 1) the dispersal of virulent isolates (probably on symptomless nursery stock) to previously uninfested production regions and 2) improved techniques for the isolation of *Phytophthora* spp., which have associated these pathogens with many declining stands previously thought to be suffering from winter injury or root asphyxiation in wet soils. The disease appears to primarily affect red raspberry and some of its hybrids, but black raspberry (*R. occidentalis* L.) cultivars may also be infected. Phytophthora root rot has not been documented on other *Rubus* spp. or hybrids.

Symptoms

Affected plants, usually found in groups, are most common where soil type, topography, or excessive rainfall or irrigation result in regular or extended periods of very high soil moisture (saturated soils). Often disease foci will first appear in low-lying sections of the planting, then expand uphill along the affected rows (Plate 42). Plant density within diseased areas is usually sparse, and the number of emerging primocanes is reduced. This is in contrast to the normal unaffected pattern of primocane emergence that generally occurs when fruiting canes have been damaged by winter injury, cane borers, or canker fungi. Both primocanes and floricanes may exhibit symptoms. Infected primocanes may rapidly wilt and collapse shortly after emergence in the spring, often after developing a dark, water-soaked lesion at their base (Plate 43). Symptom development may progress more slowly, with canes gradually turning chlorotic, wilting, and dying during the summer, or infected canes may remain symptomless until the following year. Affected floricanes typically produce weak lateral shoots with leaves that turn yellow, wilt, or scorch along the margins or between the veins. Floricanes of severely infected plants often wilt and die before harvest (Plate 44). The specific type and timing of foliar symptom expression is probably influenced by individual interactions between the raspberry genotype and the *Phytophthora* sp. and by environmental factors affecting disease initiation and development.

Scraping the epidermis from infected roots reveals a characteristic reddish brown discoloration of the necrotic cortical tissue, with a distinct margin often evident at the interface of diseased and healthy tissues (Plate 45). Necrosis sometimes extends into the crown region, in which case such margins are almost always apparent. When concurrent with the foliar symptoms, these root symptoms are usually sufficient for tentative field diagnosis of the disease.

Causal Organisms

Eight different species of *Phytophthora* have been reported as pathogens of raspberry roots. Because most are relatively slow-growing on agar media, isolation from infected tissue is greatly facilitated by the use of a selective medium for *Phytophthora* spp. All species produce hyphae that are hyaline and coenocytic but develop cross-walls with age. Generally, hyphae appear coarse and branch at obtuse angles when seen under low power of a compound microscope. Most species produce few, if any, sporangia on solid media but will abundantly produce them within one to several days if transferred to an appropriate liquid medium. Although the primary taxonomic characters of individual species are outlined below, the listed references and standard keys to the genus *Phytophthora* should be consulted when attempting to identify particular isolates.

The *Phytophthora* spp. most frequently isolated from dying raspberry plants have been identified by different investigators as *P. fragariae* C. J. Hickman (eastern North America), *P. erythroseptica* Pethybr. (western North America, continental Europe), and a highly pathogenic variant of *P. megasperma* Drechs., known as type 2 (United Kingdom, continental Europe). However, recent studies have confirmed the conspecificity of these isolates and their close similarity to strawberry isolates of *P. fragariae* with regard to several morphological characters and electrophoretic patterns of mycelial proteins. Nevertheless, neither raspberry nor strawberry isolates appear to be pathogenic on the opposite host, and they may constitute separate subspecies or varieties within *P. fragariae*. Because of minor differences in morphological and cultural characters between raspberry and strawberry isolates of the fungus and apparent differences in pathogenic specialization, it has recently been suggested that raspberry isolates of the species be designated *P. fragariae* var. *rubi*.

Raspberry isolates of *P. fragariae* are characterized by their ability to form relatively abundant oospores in single culture (although apparently only on certain media), oogonia with tapered or funnel-shaped bases, predominantly amphigynous antheridia, a very slow rate of vegetative growth (with no growth at temperatures above 25–28°C), and nonpapillate sporangia regularly produced only in liquid media.

P. megasperma has been reported from the United States and the United Kingdom. Isolates belong to the Broad Host Range subgroup of the species (syn. *P. megasperma* var. *megasperma* Waterhouse) and are characterized by their ability to form abundant oospores in single culture, predominantly paragynous antheridia, spherical oogonia, inability to grow at temperatures above 30°C, and nonpapillate sporangia produced only in liquid media.

P. cactorum (Lebert & Cohn) J. Schröt. has been reported from the United States and the United Kingdom. The species is homothallic, and both oospores and sporangia are abundantly formed on solid media. Sporangia are markedly papillate, borne on sympodially branched sporangiophores, and deciduous and do not proliferate after discharging. Antheridia are paragynous.

P. citricola Sawada has been reported from the United States and the United Kingdom. The species is homothallic, with paragynously attached antheridia. Sporangia regularly form only in liquid media. They are persistent, semipapillate (often with two distinct apices), and borne on sympodially branched sporangiophores, and they do not proliferate after discharging. Colonies produce a distinctly petaloid pattern on many agar media.

P. cryptogea Pethybr. & Lafferty has been reported from the United States and Australia. As presently defined, it is a heterogeneous taxon encompassing a range of morphologically related types. All are heterothallic; thus, isolates will normally not form oospores unless paired with another isolate of the opposite mating type (whereupon antheridium attachment is amphigynous). The nonpapillate sporangia are rounded at the base and proliferous. They form on simple, undifferentiated sporangiophores primarily in liquid media, although a limited number may also form on solid. Some isolates also form spherical hyphal swellings in liquid media.

P. drechsleri Tucker has been reported only from the United Kingdom. It is morphologically similar to *P. cryptogea*, differing primarily in its ability to grow at 36°C and in the shape of its sporangia, which may be tapered at the base.

P. cambivora (Petri) Buisman has been reported only from the United Kingdom. Like *P. cryptogea* and *P. drechsleri*, it is heterothallic and produces nonpapillate, proliferating sporangia on simple sporangiophores. However, it is readily distinguishable from other similar species by the morphology of its gametangia, which are produced when grown in dual culture with a *Phytophthora* sp. of the opposite mating type. Oogonial walls are ornamented with wartlike bullate protuberances, and the amphigynously attached antheridia are two-celled.

An indeterminate number of unidentified *Phytophthora* spp. have also been isolated from raspberry roots in the United States, the United Kingdom, and Australia, although pathogenicity has been reported only for those from the United States.

Disease Cycle and Epidemiology

Many details of the disease cycle must be surmised from circumstantial evidence and by extrapolating experimental data acquired from studies of other *Phytophthora* spp. on different hosts.

There are several potential sources of primary inoculum, but their relative importance almost certainly varies among species. For instance, some species (such as *P. megasperma* and *P. cactorum*) have wide host ranges, are very widely distributed, and are likely to be present in many soils prior to the establishment of new raspberry plantings. In contrast, raspberry isolates of *P. fragariae* have an extremely narrow host range, are less widely distributed than many other *Phytophthora* spp. that attack raspberries, and are probably introduced into most individual planting sites. Common means of introducing this and other *Phytophthora* spp. include the use of infected planting material, the movement of infested soil or runoff water from nearby infested sites, and the use of infested surface water for irrigation.

Once present, *Phytophthora* spp. persist primarily as mycelium in recently infected tissue or as dormant oospores that are liberated into the soil when infected tissues die and decompose. Oospores may remain viable for a number of years in the absence of a host and are relatively insensitive to environmental extremes and fungicides applied to the soil. The extent of oospore production by heterothallic species in nature is not known.

The most common inoculum is the biflagellate zoospore, produced within sporangia, which arise from the mycelium within infected tissues or from germinating oospores. For individual *Phytophthora* spp., the influence of temperature on the production of sporangia is very similar to the influence on vegetative growth (although the optimum and maximum temperatures may be slightly lower). Similarly, the indirect germination of sporangia (i.e., zoospore release) is often inhibited at the higher end of the temperature range for individual species. Approximate temperature ranges for the production and release of zoospores for the species commonly associated with raspberries are 4–25°C or less and an optimum of 13–19°C for *P. fragariae*, ≤4–25°C, optimum 10–19°C for *P. megasperma*, and ≤4–28°C, optimum 7–22°C for isolates of *P. cryptogea* from eastern North America. In general, the production of sporangia appears to be favored in very wet to moist soils (although precise requirements appear to vary among *Phytophthora* spp.) and may be further influenced by the source from which the sporangia originate.

Soil moisture requirements for discharge and dispersal of zoospores are stringent and probably the primary reason for the consistent association of Phytophthora root rot with wet soils. Optimum discharge of zoospores occurs when soils become completely saturated with water and is severely impeded as soon as the soil begins to drain. Relatively few zoospores are released at a soil water matric potential of −10 mbar (i.e., the equilibrium value in soil 10 cm above a water table), and virtually no zoospores are released at matric potential values below −25 mbar. At optimum temperatures, zoospore discharge begins within approximately 30–60 min after soils became saturated and is largely completed within an additional 3–6 hr, provided the soil remains saturated. The influence of soil saturation on zoospore dispersal is also very important. Significant movement of zoospores (flagellar propulsion) through the soil occurs only when the largest pores in the soil are saturated with water. Flowing or splashing water on the surface of flooded soils also provides an efficient means for the dispersal of passive zoospores. Furthermore, it is likely that periods of waterlogging may cause root cracking and an increased release of root exudates, to which zoospores are chemotactically attracted.

Because episodes of soil saturation or near saturation are required for significant production and dispersal of infective propagules, they may constitute infection periods for initiating Phytophthora root rot. The necessary duration of these flooding episodes for infection to occur is influenced by interactions among the *Phytophthora* spp. present, pathogen population, soil temperature, and host genotype. For instance, when plants are grown in soil infested with *P. fragariae* under greenhouse conditions, highly susceptible cultivars are killed if the soil is watered daily with unimpeded drainage (matric potential of 0 to −10 mbar); moderately susceptible cultivars are not killed until a single flooding episode of 8–24 hr occurs; and relatively resistant cultivars become infected but remain alive following repeated 24-hr flooding episodes. In contrast, *P. megasperma* is generally nonpathogenic in the absence of prolonged flooding periods but causes severe disease or death in many cultivars following regular 48-hr flooding episodes. In addition to its effects on pathogen biology, waterlogging may also contribute to disease development by interfering with oxygen-dependent host defense mechanisms (a hypothesis supported by limited experimental evidence).

Once infection occurs, the fungus may produce new sporangia and zoospores capable of causing further infections whenever soil moisture and temperature conditions are favorable.

Control

Control programs should integrate as many of the following components as practical.

Exclusion. Several *Phytophthora* spp. pathogenic on raspberries are apparently not common soil inhabitants and therefore, in theory, may be excluded from sites on which they do not occur. If possible, planting material that has not been propagated or grown in field soil (greenhouse-propagated plants from meristem tip culture) should be used. Infected or infested planting stock is strongly suspected to be an important means of disseminating *Phytophthora* on raspberry. In addition, the movement of soil or water from symptomatic to asymptomatic plantings should be avoided.

Site selection and modification. Rapid drainage of excess water from the plant root zone is of paramount importance in controlling *Phytophthora* root rot. Sites with soils that are slow to drain or low-lying areas where water may periodically accumulate should be avoided. A raised-bed planting system or the installation of drainage tile may minimize the duration of saturation on marginal sites.

Species and cultivar selection. Phytophthora root rot is most serious on red raspberry and some of its hybrids. The black raspberry cultivars Cumberland and Munger are reported to be susceptible to *P. fragariae*, and the cultivars Bristol, Dundee, and Jewel appear to be moderately to highly resistant, both in controlled studies and in the field in eastern North America.

Among red raspberry cultivars recommended for commercial production in northwestern North America, Meeker, Chilliwack, and Sumner show some degree of field resistance to *P. fragariae*, but Willamette, Skeena, Chilcotin, and Comox are all highly susceptible. Among cultivars grown in eastern North America, Titan and Hilton are extremely susceptible; Festival, Heritage, Reveille, and Taylor, moderately to highly susceptible; Newburgh, somewhat resistant; and Latham, significantly resistant. All popular cultivars from the United Kingdom breeding program appear to be highly susceptible. Apparently, cultivars derived primarily from the North American *R. idaeus* L. subsp. *strigosus* (Michx.) Maxim. are generally more resistant than those from the European *R. idaeus* subsp. *vulgatus* Arrhen.

The extent to which resistance to *P. fragariae* correlates with resistance to other *Phytophthora* spp. has not been widely investigated; however, some red and black raspberry cultivars resistant to *P. fragariae* in controlled tests are susceptible to *P. megasperma* in the same tests.

Chemical control. Acylalanine (metalaxyl) and phosphorous acid (fosetyl Al) have provided significant levels of control in a number of regions where their use is allowed. However, these materials often fail to provide adequate control when used on moderately to highly susceptible cultivars in wet soils and thus should be used only as one component of an integrated disease management program. In order to maintain the viability of a chemical control option, these fungicides should never be used in propagation beds—otherwise, resistant strains of the pathogen might develop and be distributed with the planting stock. Soil fumigants are considered to be effective in reducing oospore populations, but a mere reduction of primary inoculum is probably insufficient to control Phytophthora root rot.

Selected References

Barritt, B. H., Crandall, P. C., and Bristow, P. R. 1979. Breeding for root rot resistance in red raspberry. J. Am. Soc. Hortic. Sci. 104:92-94.

Bristow, P. R., Daubeny, H. A., Sjulin, T. M., Pepin, H. S., Nestby, R., and Windom, G. E. 1988. Evaluation of *Rubus* germplasm for reaction to root rot caused by *Phytophthora erythroseptica*. J. Am. Soc. Hortic. Sci. 113:588-591.

Duncan, J. M., and Kennedy, D. M. 1989. The effect of waterlogging on Phytophthora root rot of red raspberry. Plant Pathol. 38:161-168.

Duncan, J. M., Kennedy, D. M., and Seemüller, E. 1987. Identities and pathogenicities of *Phytophthora* spp. causing root rot of red raspberry. Plant Pathol. 36:276-289.

Wilcox, W. F. 1989. Identity, virulence, and isolation frequency of seven *Phytophthora* spp. causing root rot of raspberry in New York. Phytopathology 79:93-101.

(Prepared by W. F. Wilcox)

Verticillium Wilt

Verticillium wilt is caused by a common soilborne fungus that has been reported to cause wilt on more than 160 different kinds of plants, including strawberries, eggplant, tomatoes, potatoes, stone fruits, and peppers. It is a serious disease of black raspberries and is often called *bluestem* or *blue stripe wilt*; it also occurs on red and purple raspberries and blackberries. It is most common in the northern half of the United States and along the Pacific coast, particularly in California.

Verticillium wilt can be widespread and extremely destructive. Losses are generally greatest in black raspberries and susceptible cultivars of blackberry. Economic losses occur from reduced yields caused by wilting, stunting, and death of the fruiting cane or the entire plant. The disease is less severe in red raspberry; infected plants may survive for years but produce fewer primocanes and have reduced vigor and berries that may be small and crumbly. The disease has been reported in Europe but is generally not considered a serious problem.

Symptoms

On black and purple raspberry, new infections cause leaves of new canes to turn pale in midsummer. Plants may appear to recover during cool fall weather. In the following spring, leaves on infected fruiting canes may turn yellow (beginning at the bottom and moving upward), and then wilt and die. Infected canes are often stunted and may turn entirely blue or blue on one side before they wither and die (Plate 46). The bluish tinge imparted to the cane accounts for the name *bluestem*. Infected vessels in the vascular tissue generally have a red discoloration. Infected plants usually die in 1-3 years.

Symptoms on red raspberry tend to be less severe than on black raspberry, but leaf symptoms are similar. Leaflets on infected plants often fall before the petioles drop, and a tuft of leaves may remain at the tip of infected canes. Cane discoloration is not as evident as on black raspberry. The plants may survive for years, but the amount of suckering and general plant vigor may be greatly reduced.

On blackberries the infected canes wilt, and the leaves turn yellow and become brown and necrotic. The advent of cool weather in the fall may bring about a virtual disappearance of symptoms. Canes do not turn blue, as they do in infected raspberry. Infected fruiting canes that survive the winter often leaf out and set fruit; however, as the fruit ripens in warmer weather, the canes usually collapse.

Verticillium wilt may affect only part of a raspberry plant if the fungus has not invaded all the roots, so part of the plant may be dying while the rest appears to be healthy.

Field symptoms are not always a reliable indication of the disease. Other root diseases, insect feeding, wind damage, and excess soil moisture can produce similar symptoms.

Causal Organism

There is some confusion in the literature about the identity of the fungus causing Verticillium wilt of *Rubus*; both *Verticillium albo-atrum* Reinke & Berthier and *V. dahliae* Kleb. are reported as causal organisms. This is partly because of past disagreements among researchers about whether these two fungi are distinct species or biotypes of a single species. Prior to the early 1970s, both were considered forms of *V. albo-atrum*, but most authorities now recognize them as separate. Strain variation is also considerable within both species, and not all strains can cause wilt disease in *Rubus* spp.

Conidiophores of *V. albo-atrum* are 100–300 μm long (or longer) and develop whorls of branches (Fig. 22). There are one to eight primary whorls per conidiophore, usually 30–90 μm apart, which sometimes bear secondary whorls with straight or slightly bowed branches 13–38 μm long. Conidia are borne singly on phialides but may often collect into heads. Conidia (4.0–11 × 1.7–4.2 μm) are ellipsoidal and unicellular. The mycelium is septate and hyaline, turning brown with age. *V. dahliae* is identical to *V. albo-atrum* in these characteristics. The two species are separated by the ability of *V. dahliae* to form true microsclerotia as survival structures; *V. albo-atrum* forms only melanized hyphae within infected tissues. They also differ in temperature sensitivity: *V. albo-atrum* grows optimally at temperatures up to 24°C, and *V. dahliae* grows well up to 27°C.

Disease Cycle and Epidemiology

Much of the following information is drawn from studies of other crops to supplement the more limited information available concerning Verticillium wilt of *Rubus*. The fungi survive as microsclerotia or melanized hyphal fragments bound within plant debris or free in the soil. Hyphae from these structures penetrate root hairs or the root cortex directly and enter xylem vessels. The fungi can also penetrate through breaks or wounds in the roots. Movement within the host is accomplished by growth of hyphae through the vessels and by movement of conidia in the transpiration stream. Upon the death of infected plant parts, these fungi form new survival structures and are returned to the soil, where they survive for long periods. *V. albo-atrum* occurs primarily in the top 30 cm of soil but has been recovered from depths of 1 m. It has been reported to survive in soil for up to 14 years in California in the absence of known hosts.

Disease severity is affected by the population of the fungus in the soil, which in turn is affected by previous crop history. Raspberries appear to be most susceptible during the first few years of growth. Other factors reported to increase disease incidence and severity include low temperatures during the previous winter, heavy soils, and cold, wet spring weather. Symptoms often become acute in midsummer because of the combined effect of damage to vascular tissues by the fungus and the moisture stress associated with hot, dry weather.

Control

Applications of fungicides are ineffective in control. Soil fumigation has provided excellent control in some locations but is generally very expensive. Reintroduction of the pathogen into fumigated soils, accompanied by a consequential buildup in population, is a major concern with this practice. Rotations (3- to 4-year) with nonsusceptible crops have been recommended for control in Canada but were not effective in California.

Only disease-free nursery stock from fields known to be free of *Verticillium* should be used to establish new plantings.

Satisfactory resistance in commercial raspberry cultivars is not available. It is generally recommended that raspberries not be replanted in an area where the disease has been a problem. If they are replanted in an infested site, soil fumigation should be considered.

Selected References

Harris, R. V. 1925. The blue stripe wilt of raspberry. J. Pomol. Hortic. Sci. 4:221-229.

Hawksworth, D. L., and Talboys, P. W. 1970. *Verticillium albo-atrum*. Descriptions of Pathogenic Fungi and Bacteria, No. 255. Commonwealth Mycological Institute, Kew, Surrey, England.

Hawksworth, D. L., and Talboys, P. W. 1970. *Verticillium dahliae*. Descriptions of Pathogenic Fungi and Bacteria, No. 256. Commonwealth Mycological Institute, Kew, Surrey, England.

Wilhelm, S. 1955. Longevity of the Verticillium wilt fungus in the laboratory and field. Phytopathology 45:180-181.

Wilhelm, S., Storkan, R. C., and Sagen, J. E. 1961. Verticillium wilt of strawberry controlled by fumigation of soil with chloropicrin and chloropicrin-methyl bromide mixtures. Phytopathology 51:744-748.

(Prepared by M. A. Ellis)

Armillaria Root Rot

Armillaria root rot is an economically important disease in a wide range of woody crop plants worldwide. Although *Armillaria* spp. are probably best known as pathogens of forest and orchard crops, they have been recorded attacking *Rubus* in North America, Europe, Australia, and New Zealand. While not common, attacks in commercial *Rubus* plantings can nevertheless cause serious losses, since the disease is generally fatal and difficult to control. Besides inhabiting the roots of infected trees and shrubs, *Armillaria* spp. can persist for decades as wood-decaying fungi in tree stumps and other woody debris in the ground. Since these fungi can spread underground from such material, there is a risk of Armillaria root rot in crops sited on any land cleared of trees or adjacent to existing trees.

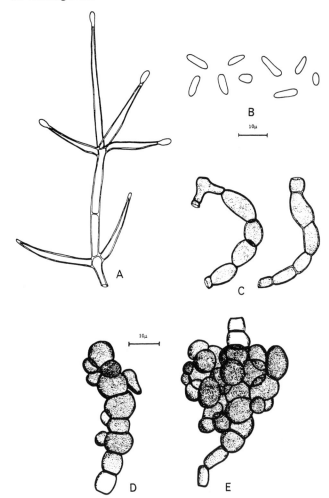

Fig. 22. Verticillate conidiophore (**A**), conidia (**B**), and dark, resting mycelium (**C**) of *Verticillium albo-atrum*; young microsclerotia (**D**) and mature microsclerotia (**E**) of *V. dahliae*. (Reprinted, by permission, from Hawksworth and Talboys, 1970)

Armillaria spp. are collectively known by several common names, of which the most frequently used is probably *honey fungus*. In the United States, the names *oak fungus* and *oak root fungus* have often been used because of the common occurrence of the disease on sites containing oak stumps.

Symptoms

The most obvious symptoms of Armillaria root rot are common to many fatal root diseases: top growth declines and dies, as the roots and crown of the plant are progressively invaded and killed (Plate 47). This process can take a few weeks or many months, depending on the resistance offered by the plant and the pathogenicity of the *Armillaria* spp. involved.

Because of the way in which the fungus spreads underground, death of plants usually occurs in patches around sources of inoculum. At first only a few scattered plants may be affected, but without intervention more are likely to die, and eventually a sizeable group of dead plants may develop.

On individual infected plants, the fungus is generally visible as white or yellowish white sheets of fungal mycelium below dead or dying bark on the main roots and crown (Plate 48). These mycelial sheets are about as thick as paper and often grow in fanlike patterns or have fanlike striations on their surface. Other fungi may produce similar mycelium in long-dead plant material, but few others are likely to produce such substantial sheets in living or recently killed plants.

The presence of mycelial sheets is the most reliable sign of *Armillaria* infection, but many species also produce rhizomorphs. These structures, the means by which the fungus spreads through the soil, can be a valuable aid to diagnosis if they occur in the vicinity of dead and dying plants. They are branching, rootlike strands, up to 3–4 mm in diameter, which consist of a core of white mycelium surrounded by a dark red to blackish rind. They may be found attached to or ramifying among the roots of suspect plants. The prolific, tough, black rhizomorphs characteristic of some North American and European species have given rise to the common names *bootlace* and *shoestring fungus*, which are sometimes applied to *Armillaria*.

In autumn, the mushroom fruiting bodies of the fungus may appear around the bases of infected plants. These generally have a yellowish brown cap (hence the common name *honey fungus*), and most species have a conspicuous ring around the stem just below the cap. *Armillaria* mushrooms can easily be confused with those of other fungi; diagnosis of Armillaria root rot, without expert guidance, should not rest on the presence of these mushrooms alone.

Some *Armillaria* spp. are common secondary invaders of plants weakened or killed by other agents. This behavior should be considered when trying to assess the importance of the fungus as a cause of disease in any particular situation. The presence of mycelium in the roots or collar of still partially live plants is probably the best indicator that *Armillaria* is causing the problem.

Causal Organisms

Armillaria root rot used to be attributed to a single fungus, *A. mellea* (Vahl:Fr.) P. Kumm., but it is now recognized that several species are alike in appearance but differ in pathogenicity and ecology. The species that are considered major tree pathogens in North America and Europe are *A. tabescens* (Scop.) Dennis, Orton, & Hora, *A. ostoyae* (Romagnesi) Herink, and *A. mellea*; in Australia and New Zealand they are *A. novae-zelandiae* (Stevenson) Herink, *A. luteobubalina* Watling & Kile, and *A. limonea* (Stevenson) Herink.

There is no information to link any particular *Armillaria* spp. directly with disease in *Rubus*, but any of the pathogens named above are likely to be capable of killing plants. It is also possible that small hosts, such as cultivated *Rubus* spp., could be at risk from other *Armillaria* spp., which are regarded as weak and relatively harmless pathogens in forest crops.

Disease Cycle and Epidemiology

Armillaria attacks invariably arise by vegetative underground spread of the pathogen from infected woody material. Although cases are known in which the fungus was introduced to sites in river-borne forest debris or in wood used for fence posts or drainage pipes, the initial sources of infection in commercial *Rubus* plantings are most likely to be either tree stumps remaining after initial clearing of the land or stumps and infected trees in adjacent woods or hedgerows. The fungus can spread by mycelial growth through roots it has killed or spread through the soil as rhizomorphs. New infections are initiated by rhizomorphs or the transfer of mycelium at contacts between roots. The relative importance of rhizomorphs for spread and infection varies with the species of *Armillaria* and site conditions. A very few species, such as *A. tabescens* in the southeastern United States, are not known to produce rhizomorphs under any field conditions. In a few others, such as *A. luteobubalina* in Australia, rhizomorphs may be sparse or absent under some soil conditions. Otherwise, in North America, Europe, Australia, and New Zealand rhizomorphs are an important means of spread and the predominant means of infection.

Control

Control of Armillaria root rot can be achieved by removing or isolating infected woody material. Infected plants, together with the immediately adjacent, apparently healthy plants, should be completely grubbed out and either removed from the site or destroyed. In addition, it is extremely important to identify and attend to the initial source of inoculum. The root systems of canes, when compared to the initial source, are likely to represent a small and short-lived, though not negligible, reservoir of inoculum for further infection. If the source of inoculum is a stump within the planting, it should be removed as thoroughly as possible. These measures are unlikely to eradicate the fungus, because some infected root fragments will inevitably remain in the soil. Nevertheless, an economically satisfactory level of control is attainable. The use of a soil fumigant combined with stump removal has proven to be, under some circumstances, a successful method of treating infected ground. However, the most effective fumigants are hazardous materials, and expert advice should be sought if this method is contemplated.

When the source of infection is on the adjacent land and cannot be destroyed, barriers to prevent rhizomorph spread and root contact between source and crop may provide effective control. Barriers can be of impervious material (such as buried plastic sheeting), ditches, or regularly cultivated strips of ground.

Selected References

Greig, B. J. W., and Strouts, R. G. 1983. Honey fungus. For. Comm. Aboricultural Leafl. 2. 12 pp.
Sinclair, W. A., Lyon, H. H., and Johnson, W. T. 1987. Diseases of Trees and Shrubs. Cornell University Press, Ithaca. 574 pp.
Smith, I. M., Dunez, J., Lelliot, R. A., Phillips, D. H., and Archer, S. A., eds. 1988. European Handbook of Plant Diseases. Blackwell Scientific Publications, Oxford. 583 pp.

(Prepared by S. C. Gregory)

White Root Rot

White root rot, caused by a species of *Vararia* (Lachnocladiaceae), is a sporadic disease of raspberry that has been reported only from Australia. The disease causes a root rot, dieback, and death of plants. A white root rot of raspberry has also been reported from New Zealand but is apparently caused by unrelated fungi.

The disease has been reported on field-infected raspberry, on loganberry, and once on plum. Inoculation studies have shown that other rosaceous plants can be infected, but raspberry is the only host of economic importance. In some cases, white root rot has virtually destroyed entire plantings over 3–4 years. More commonly, only a few plants are killed annually.

White root rot causes yellowing, wilting, and dieback of canes, which is most rapid in young plants. Roots and crowns are also rotted and characteristically covered with a dense, white mycelial mat (Plate 49). White rhizomorphs may develop on the surface of these parts.

The basidiospores ($10–15 \times 2.3–3.0$ μm) of the fungus, which are not readily found on infected material, are elliptic-fusiform, hyaline, smooth, and nonamyloid. Dichohyphidia (sterile, branched structures found throughout the mycelial mat) are dendriform and dichotomously branched and when mature are dextrinoid, staining red-brown in Meltzer's reagent.

The fungus requires a food base before it is capable of infecting healthy plants, and it can probably survive in soil for several years on infected roots or canes. The disease is most severe when soil moisture is low and soil temperatures are high. Spread within a planting occurs via root contact and less often by the spread of infected roots and canes through cultivation.

Control measures include adequate irrigation to prevent soils from drying out, removal and destruction of infected plants and plants immediately adjacent to them, and the fumigation of replant sites where white root rot is present.

Selected References

Pascoe, I. G., Washington, W. S., and Guy, G. 1984. White root rot of raspberry is caused by a *Vararia* species. Trans. Br. Mycol. Soc. 82:723-726.

Wade, G. C. 1951. White root rot of raspberries. Aust. J. Sci. Res. Ser. B 4:211-223.

(Prepared by W. S. Washington)

Diseases Caused by Bacteria

Crown and Cane Gall

Crown gall is a common and destructive disease that affects all *Rubus* spp. worldwide. It is caused by a bacterium that can infect herbaceous and woody plants from at least 93 families, and it can be a limiting factor in the nursery production of raspberries and blackberries. The disease produces galls that are often located at the crown of the plant at or just below the soil line (thus the name *crown gall*). However, galls may also occur on other plant parts, including roots and canes. The impact of the disease on plant growth and production can range from no apparent effect to death of the plant. Damage is greatest when young plants become infected during the year in which they are planted. Severely galled plants are weakened, stunted, and unproductive. The abnormal proliferation of plant tissues that results in gall formation disrupts water and nutrient uptake and transport in the plant.

Symptoms

The most diagnostic symptom is the formation of galls, which are generally located on the crown or roots. They may also form at pruning wounds or splits in the cane, which commonly occur from bending and training of the cane. A gall is a sphere of disorganized vascular and parenchyma tissue with a texture that is usually soft and spongy, but which can harden with age. Galls on canes vary in shape, from globose or spherical (up to 10 mm in diameter) to elongated ridges of white granular tissue (Plate 50). Small galls are easily confused with other tissue overgrowths caused by excessive callus formation at wound sites.

In temperate regions, galls generally appear with the onset of warm weather in the spring or early summer, often followed by a rapid increase in size and number. The enlargement of galls can cause canes to split open and dry out. Severely infected canes may exhibit a variety of symptoms, including stunting, foliar chlorosis, poor-quality (small and seedy) fruit, wilting, and in some cases death of the plant. Eventually gall tissues turn brown and begin to deteriorate (Plate 51), especially those in or near the soil. Although galls blacken and die during the winter, new gall tissues typically erupt the following spring in the vicinity of the old galls.

Causal Organisms

Two bacterial pathogens are associated with the crown gall syndrome on *Rubus*: *Agrobacterium tumefaciens* (E. F. Smith & Townsend) Conn and *A. rubi* (Hildebrand) Starr & Weiss. The latter has a more restricted host range and is particularly associated with galls on floricanes. Both bacteria belong to the family Rhizobiaceae. The taxonomy of *A. tumefaciens* is confusing, because it is based on the pathogenic behavior of the bacterium. This is a problem, because the genes that make the bacterium pathogenic are carried on a large extrachromosomal piece of DNA, which is referred to as the tumor-inducing (Ti) plasmid. This plasmid can be lost from pathogenic strains or transferred from pathogenic to nonpathogenic strains of the bacterium. If an *A. tumefaciens* bacterium loses the Ti plasmid, it is no longer pathogenic and cannot be distinguished from other *Agrobacterium* spp. If a nonpathogenic species such as *A. radiobacter* (Beijerinck & van Delden) Conn picks up the Ti plasmid, it then becomes pathogenic and would be classified as *A. tumefaciens*. Thus, the species of an individual bacterium can be changed according to the presence or absence of the Ti plasmid. *A. radiobacter* is often associated with strains of *A. tumefaciens* in tumors. With few exceptions, isolations from uninfected plant tissues and soil yield only *A. radiobacter*. The term *biovar* is used to described groupings of bacterial strains within a species. *A. tumefaciens* has two biovars, 1 and 2, that can be isolated from the rhizosphere and galls of *Rubus*, but disease is usually caused by biovar 2. Biovar 2 strains and *A. rubi* utilize nopaline and succinamopine, which the bacteria induce the infected host to produce and which are believed to confer nutritional advantages to the pathogens.

A. tumefaciens is rod-shaped ($0.6–1.0 \times 1.5–3.0$ μm), aerobic, and gram-negative, does not produce endospores, and is motile by means of one to six peritrichous flagella. Colonies on potato-dextrose agar are usually white to cream colored, convex, and glossy and have entire margins.

Disease Cycle and Epidemiology

The pathogen requires a wound to enter the plant and initiate infection. Wounds can result from natural causes, such as lateral root formation or leaf scars, or from mechanical causes, such as pruning, training practices, cultivation, harvesting, insect feeding, or frost damage. Crown gall development is initiated when the pathogen enters a wound, attaches to a susceptible plant cell, and inserts a segment of DNA (called T-DNA) from the Ti plasmid (carried by the pathogen) into a chromosome of the healthy raspberry or blackberry cell. Expression of T-DNA results in overproduction of plant hormones. The uncontrolled synthesis of growth hormones stimulates the plant cells to divide and enlarge abnormally, forming a gall, which typically carries the pathogen. Gall development is usually accompanied by the synthesis of opines, which are unique amino acids produced in plant tissues infected by *A. tumefaciens*.

Small galls appear within 2–4 weeks after infection when temperatures are 20°C or above. Symptom development is delayed at temperatures below 15°C, and latent infections may not show signs of gall formation until higher temperatures occur during the second growing season. Infection is inhibited at temperatures above 32°C.

Once the T-DNA is incorporated into the genome of the plant, galls can continue to develop in the absence of the bacterium. This probably explains why new galls develop following the death of older ones during winter.

Control

The most effective means of control is to establish plantings in uninfested soil with pathogen-free planting material; under these conditions, the plants do not develop crown gall. Fields where crops susceptible to crown gall have been grown should be avoided for 2–3 years or more. Planting sites that have been previously planted to crops such as vegetables, grains, or pasture should be used when possible. When budded apple trees were grown in fields where previous nursery crops such as grape, peach, raspberry, and rose had been heavily infected with crown gall, the trees became severely galled. Efforts to use fumigation to eliminate *Agrobacterium* from the soil have generally been ineffective.

Sanitation and the use of good cultural practices, such as careful inspection of all nursery stock and discarding all plants with crown gall symptoms, are important in disease control. This is especially important in nurseries to avoid contamination of healthy plants. The adoption of any management practice that eliminates or minimizes wounding of plant tissues is beneficial in control. Garrett reported that wounds left after cutting out old blackberry canes remained susceptible for at least several days. These wounds were most readily infected during the growing season. Controlling root-attacking insects and nematodes reduces wounding of the roots.

Whereas *Rubus* cultivars immune to crown gall are currently unavailable, certain cultivars show more resistance than others. In the Pacific Northwest of British Columbia and the United States, the cultivar Willamette is highly resistant to the pathogen, Nootka and Canby are intermediate in resistance, and Skeena and Haida are susceptible. However, in the presence of nematodes (*Pratylenchus penetrans* (Cobb) Filipjev & Schuurmans, Steckhoven), the incidence of crown gall was significantly increased on root systems of both Skeena and Willamette. In the United Kingdom, Malling Jewel was considerably more resistant than Malling Delight with respect to growth and yield when inoculated with *A. tumefaciens* immediately before planting.

The use of a bacterial biological control agent (*A. radiobacter* strain K84) has shown promise for control of crown gall on *Rubus* and a number of other crops. This agent was effective in preventing crown gall on raspberry in Hungary and on boysenberry in field tests in Oregon. With this method, the crown and roots of plants are dipped in a suspension of *A. radiobacter* strain K84 prior to planting. This treatment provides protection against infection by *A. tumefaciens*, which is especially important during the first year of plant establishment. Biological control by strain K84 is purely preventative in nature and will not cure infections after they have occurred. This control agent is commercially available throughout North America and in several other countries worldwide.

Selected References

Brisbane, P. G., and Kerr, A. 1983. Selective media for three biovars of *Agrobacterium*. J. Appl. Bacteriol. 54:425-431.

Garrett, C. M. E. 1978. Crown gall of blackberry: Field spread and susceptibility to disease. Plant Pathol. 27:182-186.

Moore, L. W. 1980. Controlling crown gall with biological antagonists. Am. Nurseryman 151:40, 42, 44.

Moore, L. W., and Cooksey, D. C. 1981. Biology of *Agrobacterium tumefaciens*: Plant interactions. Pages 15-46 in: The Biology of the Rhizobiaceae. (Suppl. Int. Rev. Cytol.) K. L. Giles and A. G. Atherly, eds. Academic Press, New York.

Ophel, K., and Kerr, A. 1990. *Agrobacterium vitis* sp. nov. for strains of *Agrobacterium* biovar 3 from grapevines. Int. J. Syst. Bacteriol. 40:236-241.

Swait, A. A. J. 1980. Field observations on disease susceptibility, yield and agronomic character of some new raspberry cultivars. J. Hortic. Sci. 55:133-137.

Vrain, T. C., and Copeman, R. J. 1987. Interactions between *Agrobacterium tumefaciens* and *Pratylenchus penetrans* in the roots of two red raspberry cultivars. Can. J. Plant Pathol. 9:236-240.

(Prepared by L. W. Moore)

Fire Blight

Fire blight is a common and very serious disease of many rosaceous plants, including pear and apple, but on *Rubus* cane fruits it is relatively uncommon and rarely of economic importance. It has occasionally been reported from Illinois, Maine, and North Carolina. More recent outbreaks occurred in Ohio, Wisconsin, and Illinois in the 1980s and in Wisconsin in 1990. The disease attacks red raspberry, wild blackberry, and cultivars of thornless semierect and thorny erect blackberry. Disease losses result from fruit infection and less commonly from death of tips of primocanes.

Symptoms

Fire blight infections may initially appear at the cane tip, where infection may proceed downward for distances up to 20 cm; at axillary buds, causing cane girdling and death of distal parts; or at flower and fruit clusters. The lesions are water-soaked and produce abundant bacterial ooze. Diseased portions of canes become necrotic and purplish black, and the tips of primocanes may become curved in the form of a shepherd's-crook (Plate 52). Infected berries do not mature; they become brown, dry, and very hard and remain attached to the pedicel (Plate 53). Entire fruit clusters can be infected, but generally a few berries in each cluster remain healthy. Fruit losses of 65% have been observed on thornless blackberry cultivars in Illinois.

Causal Organism

Erwinia amylovora (Burr.) Winslow et al, the cause of fire blight, is a facultatively anaerobic, motile, gram-negative, rod-shaped bacterium approximately 0.5×2.0 μm. As is typical of bacteria, it has a short generation time of 1–2 hr at 25°C. The bacterium may be cultured on a variety of simple organic media, including nutrient agar, nutrient-yeast-dextrose agar, or modified Emerson's medium. Several selective media have been developed, including that of Miller and Schroth and that of Crosse and Goodman. They are useful for isolating the bacterium from infected plant tissue or epiphytic populations from plant surfaces.

Disease Cycle and Epidemiology

The disease cycle of fire blight on *Rubus* has never been studied but probably has similarities to the disease on other hosts. *E. amylovora* spends its entire life cycle in association with tissues of living hosts. The initial inoculum source each spring is probably from overwintering cankers on *Rubus* spp. Strains of the bacterium from *Rubus* are unique. Unlike isolates from other rosaceous plants, isolates from *Rubus* infect only *Rubus*; apple and pear isolates are not pathogenic on *Rubus*. Bacterial populations increase early in the spring on cankers and move either with prevailing winds in rainstorms or by insects to flowering *Rubus* cane fruits. Warm temperatures (18–25°C) and light rain favor infections. Flowers, fruit, cane tips, and succulent lateral shoots become infected, initially appearing water-soaked but eventually turning brown (on fruit) or black (on shoots and cane tips).

Control

No control measures have been developed or are generally warranted, considering the sporadic nature of the disease on *Rubus*. However, removal of infected tissues by pruning and encouraging air movement to facilitate rapid drying of susceptible plant parts may reduce the likelihood of infection.

Selected References

Crosse, J. E., and Goodman, R. N. 1973. A selective medium for and a definitive colony characteristic of *Erwinia amylovora*. Phytopathology 63:1425-1426.

Folsom, D. 1947. Bacterial twig and blossom blight of raspberry in Maine. Plant Dis. Rep. 31:324.

Lelliot, R. A. 1974. *Erwinia*. Pages 332-430 in: Bergey's Manual of Determinative Bacteriology. 8th ed. William & Wilkins, Baltimore.

Ries, S. M., and Otterbacher, A. G. 1977. Occurrence of fire blight on thornless blackberries in Illinois. Plant Dis. Rep. 61:232-235.

Starr, M. P., Cardona, C., and Folsom, D. 1951. Bacterial fire blight of raspberry. Phytopathology 41:915-919.

(Prepared by S. M. Ries)

Pseudomonas Blight

Pseudomonas blight is a relatively minor disease of red raspberry, occurring in western Oregon and Washington State in the United States, the Edmonton area of Alberta, and the southwest corner of British Columbia. A similar disease, described as *halo blight*, has been reported in Yugoslavia on cultivated and wild red raspberry.

Symptoms

Symptoms appear in early April as brown, water-soaked spots on leaves, petioles, internodes, and tips of primocanes and developing laterals (Plate 54). The spots enlarge and blacken, and brown to black streaks extend from the blackened tips into the vascular tissues. Entire laterals often blacken and die. New growth may be killed to ground level. There is evidence that buds weakened by spur blight may be further weakened by Pseudomonas blight and fail to open. This type of damage usually ceases about mid-May. Although this disease does not produce the copious amounts of ooze typical of many similar bacterial diseases, thin strands of the substance can occasionally be seen on the surface of infected tissues. This glistening ooze is fairly diagnostic of the disease, but it is only visible during extended dry periods after infection has occurred and is often difficult to detect.

Symptoms can also appear in the fall as leaf spotting and black streaking of the cambium layer under the bark, especially in fields that are still actively growing as a result of excessive nitrogen usage, early topping of canes, or resumption of growth after summer drought. These symptoms are often difficult to detect and are easily missed.

Pseudomonas blight may be easily confused with a number of other diseases and disorders of red raspberry. Leaves affected by spur blight (caused by *Didymella applanata* (Niessl) Sacc.) appear similar to those affected by bacterial blight in early stages but do not blacken, and lesions are confined by the vascular system, which gives the lesion an angular appearance not evident in bacterial blight. The raspberry cane borer causes the blackened, wilted cane tips symptomatic of bacterial blight, but borer-infested canes break off easily, and a few cuts with a knife will show the presence or absence of the larva and its tunnel. Certain types of herbicide injury, winter injury, and root damage from root-rotting pathogens or high nematode populations can also result in symptoms that may be confused with bacterial blight. The only way the disease can be positively diagnosed is by isolation and identification of the bacterium.

Causal Organism

Pseudomonas syringae van Hall, the causal agent of Pseudomonas blight, is a gram-negative, non-spore-forming rod. The bacterium is oxidase- and arginine dihydrolase-negative and unable to pit polypectate gel, produces fluorescent pigment on King's medium B, is able to utilize lactate, produces levan when grown on a sucrose medium, produces the toxin syringomycin, and is positive for the tobacco hypersensitivity reaction. Some isolates are unable to produce the pigment or the toxin and to utilize lactate. The organism has not been characterized to pathovar status. Its pathogenicity to red raspberry is variable, depending in part on the ice nucleation abilities of the isolate.

Disease Cycle and Epidemiology

The bacterium overwinters in buds and as an epiphyte on the canes of the raspberry plant. In early spring, the bacterium functions as an ice nucleator, raising the temperature at which water freezes within plant tissues. Ice nucleation does not occur unless the population of bacteria is relatively high. The bacterium subsequently invades the tissue damaged by freezing. Not all strains of *P. syringae* are equally efficient in their ability to cause ice nucleation, and the population of bacteria required for this effect varies among strains. Some strains are unable to cause it at any population. Although wounds resulting from ice nucleation damage are the most likely infection sites, infection may also occur through any type of wound.

Cool weather and moisture favor development of the disease, but infection generally does not take place unless there has first been ice nucleation damage. This helps to explain the rather erratic occurrences of the disease.

The bacteria become systemic within the plant tissue, and infection continues to spread as long as the weather remains cool and wet. As the temperature rises and it becomes drier during late spring or summer, the disease usually subsides until cooler weather occurs again in the fall. Infections in fall take place in tissue damaged mainly by physical factors, such as windblown soil particles, wind rub, insects, and a variety of cultural practices.

Control

Some control of Pseudomonas blight can be obtained by reducing the populations of bacteria on the canes and buds with a delayed dormant copper spray of Bordeaux mixture (12-12-100). No satisfactory chemical control measure has been developed for fall infections, but an application of Bordeaux mixture before the fall rains may be beneficial.

Excessive application of nitrogen favors the growth of succulent tissues, which are most susceptible to infection. Fertilization should be based on soil or foliar analysis, and the use of excessive nitrogen should be avoided.

The cultivars Chilcotin, Newburgh, and Nova show a high level of resistance. Nootka is highly susceptible, and Skeena and Chilliwack are moderately susceptible. Meeker, Willamette, Comox, Haida, and Centennial are intermediate in reaction. Earlier observations indicated Newburgh and Viking were resistant, and Willamette, Rideau, and Carnival exhibited intermediate resistance. Sumner, Matsqui, and Fairview showed some resistance but did not rate as high as the intermediate group.

Selected References

Lelliot, R. A., Billings, E., and Hayward, A. C. 1966. A determinative scheme for the fluorescent plant pathogenic pseudomonads. J. Appl. Bacteriol. 29:470-489.

Ormrod, D. J. 1973. Bacterial blight believed responsible in raspberry dead bud. Plant Pathol. Notes. B.C. Dep. Agric. No. 15. p. 1.

Pepin, H. S., Daubeny, H. A., and Carne, I. C. 1967. Pseudomonas blight of raspberry. Phytopathology 57:929-931.

Ranković, M., and Šutic, D. 1973. Etioloska proucavanja orelne pegarosti nekih sort maline (The etiology of halo blight in certain raspberry cultivars). (In Serbo-Croatian, with English and French summaries.) Zast. Bilja 24:311-316.

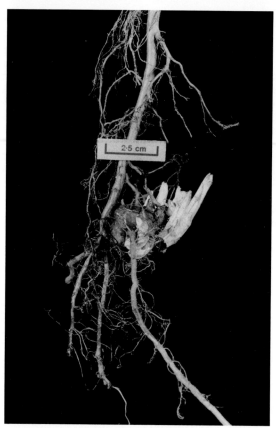

Fig. 23. Symptoms of leafy gall, caused by *Corynebacterium fascians*, on red raspberry roots. (Courtesy M. C. M. Perombelon; copyright Scottish Crop Research Institute. Used by permission)

Sinnott, N. M. 1975. Etiology and epidemiology of bacterial blight of red raspberry in British Columbia. M.S. thesis. University of British Columbia, Vancouver.

(Prepared by H. S. Pepin)

Leafy Gall

Corynebacterium fascians (Tilford) Dowson causes fasciation of sweet pea and leafy gall of chrysanthemum and several other plants. Leafy galls caused by *C. fascians* have been reported in red raspberry nurseries planted in infested wet soils. The pathogen induces the development of many buds or short, thick, distorted stems with misshapen leaves at soil level (Fig. 23). Leafy galls may superficially resemble, and can be confused with, crown and root galls caused by *Agrobacterium tumefaciens* (E. F. Smith & Townsend) Conn.

The bacterium is an aerobic, gram-positive, non-sporeforming, and nonmotile rod (0.5–0.9 × 1.5–4 μm). It is soilborne and may be transmitted on plant cuttings or nursery stock. The disease is rare, and specific control measures are generally not required.

Selected Reference

Bradbury, J. F. 1967. *Corynebacterium fascians*. Descriptions of Pathogenic Fungi and Bacteria, No. 121. Commonwealth Mycological Institute, Kew, Surrey, England.

(Prepared by M. A. Ellis)

Hairy Root

Agrobacterium rhizogenes (Riker et al) Conn is the causal agent of hairy root disease on stems and roots of a wide range of plants. The disease has been reported on *Rubus* spp. but is generally considered rare. Symptoms on apple consist of a very large number of small roots protruding from stems or roots or from localized hard swellings at graft unions. On roses these structures protrude from the ends of cuttings or at disbud scars. These malformations may be fleshy at first but eventually become fibrous. The disease is distinguished from crown gall, caused by *A. tumefaciens* (E. F. Smith & Townsend) Conn, by the production of fibrous roots on the surface of the gall.

A. rhizogenes is an aerobic, gram-negative rod (0.15–0.75 × 0.55–2.59 μm). The bacterium is exclusively a wound pathogen and soilborne and may be readily transmitted with nursery stock. The disease is considered rare, and specific control measures are generally not required.

Selected Reference

Hayward, A. C., and Waterston, J. M. 1965. *Agrobacterium rhizogenes*. Descriptions of Pathogenic Fungi and Bacteria, No. 41. Commonwealth Mycological Institute, Kew, Surrey, England.

(Prepared by M. A. Ellis)

Diseases Caused by Viruses and Viruslike Agents

Worldwide, among the wild and cultivated members of the genus *Rubus*, 33 virus and viruslike diseases have been reported; of these, 15 are found in cultivars in North America. In addition, several other diseases of presumed viral nature have been described. The diseases caused by viruses and viruslike agents, including mycoplasmalike organisms (MLOs), are all transmissible by grafting or vegetative propagation, except in instances where the pathogen is unevenly distributed in the propagation material. In this compendium the major viruses and viruslike diseases of *Rubus* are grouped according to their natural vectors. Little-known or rare virus or viruslike diseases appear in a separate category, which includes discussions of viruses infecting the genus elsewhere in the world but not reported in North America.

Important recent references are cited for each disease, with special attention given to the literature published since 1980 and not covered in previous reviews of this field. More details and earlier key literature citations are given in several handbooks, looseleaf compilations, and reviews that are listed in the references below.

Insect and nematode vectors of viruses and MLOs in *Rubus* will be discussed in this section where appropriate. (For additional information and literature citations for these fauna, see Arthropod Pests.)

Selected References

Commonwealth Mycological Institute and Association of Applied Biologists. 1970–. Descriptions of Plant Viruses. Association of Applied Biologists, Institute of Horticultural Research, Wellesbourne, Warwick, England.

Converse, R. H. 1977. *Rubus* virus diseases important in the United States. HortScience 12:471-476.

Converse, R. H. 1979. Recommended virus-indexing procedures for new USDA small fruit and grape cultivars. Plant Dis. Rep. 63:848-851.

Converse, R. H., ed. 1987. Virus diseases of small fruits. U.S. Dep.

Agric. Agric. Handb. 631. 277 pp.
Jennings, D. L. 1988. Raspberries and Blackberries: Their Breeding, Diseases and Growth. Academic Press, New York. 230 pp.
Jones, A. T. 1986. Advances in the study, detection and control of viruses and virus diseases of *Rubus* with particular reference to the United Kingdom. Crop Res. 26:43-87.
Keep, E. 1989. Breeding red raspberry for resistance to diseases and pests. Plant Breed. Rev. 6:245-321.
Murant, A. F. 1974. Viruses affecting raspberry in Scotland. Scott. Hortic. Res. Inst. Assoc. Bull. 9:37-43.
Stace-Smith, R. 1984. Red raspberry virus diseases in North America. Plant Dis. 68:274-279.

(Prepared by R. H. Converse)

Aphid-Transmitted Diseases

Raspberry Mosaic Disease Complex

Raspberry mosaic disease (RMD) is caused by a complex of viruses limited to *Rubus* in their natural host range. In North America, they are all transmitted in a semipersistent manner by the large raspberry aphid (*Amphorophora agathonica* Hottes), which is limited to *Rubus*. The disease can occur in all cultivated *Rubus* spp. in North America and causes the most damage on black raspberry, less on red raspberry, and least on blackberry. Infected plants have reduced vigor, fruit yield, and quality. This disease complex is more common in eastern than western North America.

Symptoms

In North America, new infections in black raspberry, and sometimes in red raspberry under greenhouse conditions, generally cause tip necrosis, followed by severe leaf blistering, mottling (Plate 55), and plant dwarfing. Such plants are more susceptible to winter injury. Fruit yield and quality are poor. In red raspberry, cultivar-dependent mild leaf mottling, blistering, or veinbanding may occur, or infected plants may remain symptomless. Wild and cultivated blackberry and blackberry-raspberry hybrids are sometimes infected with RMD. These plants are generally tolerant and usually symptomless.

Causal Agents

The etiology of RMD in North America is complex and still imperfectly understood. Currently, five viruses worldwide are reported to participate in various combinations to cause it: black raspberry necrosis virus (BRNV), raspberry leaf mottle virus (RLMV), raspberry leaf spot virus (RLSV), an unnamed RLSV-like virus, and Rubus yellow net virus (RYNV). Some of the properties of these viruses are listed in Table 1. Only RLSV and RLMV induced diagnostic symptoms in some red raspberry cultivars in European studies. However, all are known or suspected of interacting to cause crop damage in complex with each other or with certain other *Rubus* viruses.

All members of the RMD complex can be detected by petiole insert leaflet grafting to vigorously growing, young black raspberry propagants or seedlings. In this test, shoot tip necrosis is a characteristic first symptom of BRNV, RLMV, RLSV, and the unnamed virus resembling RLSV whether they occur singly, in combination, or with RYNV. Table 1 lists the red raspberry cultivars that have been found to be differential graft hosts for RLSV and RLMV in Scotland. A rabbit polyclonal antiserum against BRNV has been prepared in Scotland. A complementary DNA probe has been prepared in Canada against the unnamed virus that resembles RLSV.

In North America when RYNV occurs alone, leaf asymmetry and veinclearing (Plate 56) are the first symptoms on grafted black raspberry. Leaf mottling, veinbanding, distortion, and plant dwarfing symptoms follow with all of these viruses. Symptoms are more severe when BRNV and RYNV are both present to produce typical North American RMD. The role of the unnamed RLSV-like virus in the RMD complex in western North America (the only area where it has been reported) has not been elucidated. Similarly, the occurrence and roles played by RLSV and RLMV in the disease in North America are unclear at this time, although these two viruses are common and important in red raspberries in Europe, where they are associated with diseases referred to as veinbanding mosaic or leaf-spotting mosaic.

Epidemiology

The viruses of the RMD complex are readily transmitted by certain species of the aphid genus *Amphorophora*. In North America, *A. agathonica* is the major aphid vector for BRNV, RYNV, and the unnamed RLS-like virus. Only one strain of *A. agathonica* is currently recognized in North America, but

TABLE 1. Component Viruses and Their Properties in the Raspberry Mosaic Virus Complex

Virus	Known in North America	Virion Morphology	dsRNA[a] Size (kbp)	Sap-Transmissibility	Easily Heat-Inactivated in Vivo	Method of Detection
Black raspberry necrosis virus	Yes	Isometric	?	Difficult	Yes	Graft to black raspberry, serology
Rubus yellow net virus	Yes	Bacilliform, 80–150 × 25–31 nm	?	None	No	Graft to black raspberry, aphid transmission
Raspberry leaf mottle virus	?	Possibly isometric	?	None	Yes	By graft: symptoms on Malling Landmark, none on Norfolk Giant
Raspberry leaf spot virus	?	?	?	None	Yes	By graft: symptoms on Norfolk Giant, none on Malling Landmark
Unnamed virus resembling raspberry leaf spot virus	Yes	?	3.7, 2.6, and 1.8	None	Yes	dsRNA band pattern and dsRNA probes

[a] Double-stranded RNA.

there is recent preliminary evidence of more than one North American strain capable of colonizing red raspberry cultivars resistant to the common strain.

In Europe, the major vector of RLSV and RLMV is *A. idaei* Börner, and four strains of this aphid are known. Several other *Amphorophora* spp. and *Aulocorthum solani* (Kalt.), *Macrosiphum euphorbiae* (Thomas), *M. fragariae* (Wlk.) (syn. *Sitobion fragariae* (Wlk.)), *Illinoia rubicola* (Oestlund) (syn. *Masonaphis* or *Oestlundia rubicola* (Oestlund)) and *Myzus ornatus* Laing are capable of transmitting various members of the raspberry mosaic virus complex in Europe and North America.

All instars of *A. agathonica* can acquire and transmit BRNV and RYNV within minutes and retain these viruses for several hours at 20°C and for several days if starved at low temperatures. There is no evidence of transovarial transmission of any of the viruses of the raspberry mosaic complex.

A. agathonica can transmit members of the raspberry mosaic virus complex from wild and cultivated *Rubus* plants (especially red and black raspberry and wineberry, *R. phoenicolasius* Maxim.) to healthy ones. Winged *A. agathonica* reach maximum numbers in New York from mid-June to mid-August and are responsible for long-distance field spread of the RMD viruses. Wingless aphids are active, drop to the ground when disturbed, and are probably involved in local movement of the viruses of the raspberry mosaic complex.

Control

There is tolerance but no known immunity to the viruses of the raspberry mosaic complex within the genus *Rubus*. Furthermore, aphicides do not commonly act quickly enough to prevent viruliferous aphids from infecting aphicide-treated plants before the insects die. Nevertheless, effective control of RMD has been achieved by the use of vector-resistant cultivars, virus-tolerant cultivars, certified planting stocks known to be virus-free, management methods to restrict the spread of RMD, and aphicides to control aphid vector populations.

Vector-resistant cultivars. Since the mid-1950s, an extensive breeding program in the United Kingdom incorporated resistance to colonization by several strains of *A. idaei*, the European aphid vector of the raspberry mosaic virus complex, into red raspberry cultivars. In a 4-year field test in Scotland, cultivars resistant to *A. idaei* became infected with viruses of the raspberry mosaic complex at a much lower rate than aphid-susceptible cultivars planted in the same field.

In 1937, North American workers noted that the old British red raspberry cultivar Lloyd George, long favored by North American plant breeders because of its desirable horticultural properties, was not colonized by *A. agathonica* (then called *A. rubi*). Lloyd George resistance to *A. agathonica* colonization has occurred in several United States cultivars with Lloyd George as a parent, including Canby and Reveille. Several other red raspberry cultivars resistant to colonization by this aphid have been released in Canada and the United States. It should soon be routine in North American red raspberry breeding programs to incorporate one or more genes for resistance to colonization by *A. agathonica* into new cultivars. There is a need for the development of similar breeding systems to incorporate aphid vector resistance into black raspberry and other commercial *Rubus* spp. in which raspberry mosaic disease is damaging.

Virus-tolerant cultivars. Not all red raspberry cultivars that support viruliferous *Amphorophora* aphids are equally damaged by RMD. Willamette, widely grown in the Pacific Northwest, is an example of an aphid-susceptible red raspberry cultivar that is rather tolerant of infection by members of the raspberry mosaic virus complex. Tolerance to the RMD viruses is heritable but complex and should be incorporated in new red raspberry selections along with vector resistance. Field observation is currently the only satisfactory way to evaluate *Rubus* selections for RMD tolerance. More rapid, reliable methods are needed in order to use this trait more efficiently in *Rubus* breeding programs.

Certified planting stocks known to be virus-free. It would be difficult to overstate the importance of using planting stock propagated from virus-free sources to reduce *Rubus* crop losses by many virus diseases, including RMD, and by other important systemic pathogens, such as downy mildew. Many state- or government-supervised *Rubus* certification programs exist worldwide, designed to produce commercial *Rubus* planting stock that is essentially free from viruses, other systemic pathogens, harmful insects, and plant-pathogenic nematodes. Few *Rubus* viruses produce characteristic symptoms in nursery plantings that would enable them to be eliminated from stock by roguing. The propagation of nursery stock from virus-tested source plants in isolation from virus reinfection by vectors is a major goal of any successful *Rubus* certification program. The temptation of many growers to purchase inexpensive, symptomless, field-run stock is strong, but experience has shown that such stock is generally infected with one or more latent viruses. When planted, it usually lacks the vigor of certified stock. It is prone to rapid decline in fruit productivity and quality, as additional viruses are vectored into the planting and interact deleteriously with those already present in the planting stock.

Management methods to restrict the spread of RMD. Decisions about where to plant, how much area to plant, and how long to retain a *Rubus* field are relevant to the control of RMD. The farther away a planting can be located from cultivated or wild *Rubus* infected with members of the raspberry mosaic virus complex, the longer the planting is likely to escape serious damage from this disease. Isolation from mosaic-infected *Rubus* is most important for black raspberries, because they are the most severely damaged by this disease. The size of a planting also affects its rate of infection by RMD. Studies in Scotland with cultivars moderately resistant to *Amphorophora* planted in various-sized square plots showed that the larger the planting, the smaller the proportion of plants exposed on the perimeter and therefore most likely to become mosaic-infected by incoming viruliferous winged aphids.

A commercial *Rubus* planting should not be kept beyond its useful economic life. If the market demands the production of fruit of a given mosaic-susceptible cultivar in a field where mosaic infection is rapid and severe, frequent replanting may be an economic necessity to maintain adequate fruit yield and quality of that cultivar.

Use of aphicides to control aphid vector populations. Aphicide spray applications are seldom effective in preventing initial infections by members of the raspberry mosaic virus complex, because of the rapidity of the infection process by viruliferous aphids. However, such sprays may be valuable in reducing the rate of secondary infections in a *Rubus* planting by controlling resident vector aphid populations. Systematic control of vector aphid populations is particularly important in *Rubus* nurseries.

When an old planting located near younger ones is removed, the old planting should first be sprayed with an effective aphicide to prevent viruliferous aphids from moving into the younger plantings. Many aphicides give very good control of *Rubus* aphids but should only be applied as permitted by the pesticide regulations currently in effect in the area of use.

Selected References

Daubeny, H. A. 1980. Red raspberry cultivar development in British Columbia with special reference to pest response and germplasm exploitation. Acta Hortic. 80:59-86.

Daubeny, H. A., and Stary, D. 1982. Identification of resistance to *Amphorophora agathonica* in the native North American red raspberry. J. Am. Soc. Hortic. Sci. 107:593-597.

Jelkmann, W., and Martin, R. R. 1988. Complementary DNA probes generated from double-stranded RNAs of a *Rubus* virus provide the potential for rapid in-vitro detection. Acta Hortic. 236:103-109.

Jones, A. T. 1981. Recent research on virus diseases of red raspberry—The viruses, their effects and control. Scott. Hortic. Res. Inst. Assoc. Bull. 19:15-25.

Jones, A. T. 1982. Distinctions between three aphid-borne latent

viruses of raspberry. Acta Hortic. 129:41-48.
Jones, A. T., and Mitchell, M. J. 1986. Propagation of black raspberry necrosis virus (BRNV) in mixed culture with Solanum nodiflorum mottle virus, and the production and use of BRNV antiserum. Ann. Appl. Biol. 109:323-336.
Keep, E. 1989. Breeding red raspberry for resistance to diseases and pests. Plant Breed. Rev. 6:245-321.
Kurppa, A., and Martin, R. R. 1986. Use of double stranded RNA for detection and identification of virus diseases of *Rubus* species. Acta Hortic. 186:51-62.
Stace-Smith, R., and Jones, A. T. 1987. *Rubus* yellow net. Pages 175-178 in: Virus Diseases of Small Fruits. R. H. Converse, ed. U.S. Dep. Agric. Agric. Handb. 631.

(Prepared by R. H. Converse)

Raspberry Leaf Curl

Raspberry leaf curl virus (RLCV) is limited to hosts in the genus *Rubus* and is persistently transmitted by the small raspberry aphid, *Aphis rubicola* Oestlund. It is found only in North America, principally in the northeastern and Rocky Mountain regions of the United States. Chronically infected raspberry plants have severely reduced yields of poor-quality fruit and are susceptible to winter injury. Infected plants have no economic value; however, catastrophic losses from this disease are rare, because RLCV spreads so slowly in infected fields.

Symptoms

On red raspberry, characteristic severe symptoms usually appear in the growing season following infection. Leaves of primocanes as well as floricanes are severely curled, distorted, and chlorotic (Plate 57). Shoots are dwarfed and may branch excessively at the nodes. Fruit on such plants is small and crumbly or seedy. Chronically infected black raspberry plants produce small, oval, greasy green leaves (Plate 58) on short, stiff, brittle, unbranched canes. Blackberries are less frequently infected than raspberries. They may either develop symptoms like those in infected red raspberry or remain symptomless, depending on the cultivar.

Causal Agents

There are two strains of RLCV—the alpha strain, which infects red raspberry, and the beta strain, which infects red and black raspberry; both infect blackberry. The two strains do not cross-protect against each other. Both are readily transmitted by leaf or inarch (approach) grafting, and vector transmission is of a persistent type by the aphid *A. rubicola*. Nothing is known about the morphology, biochemistry, serology, or physical properties of these viruses. They have not been successfully sap-transmitted.

The routine method of detecting RLCV in suspect symptomless plants is leaf-grafting to red raspberry or to *R. phoenicolasius* Maxim., which is very sensitive to RCLV.

Epidemiology

A. rubicola, the only known vector of both RLCV strains, is a sluggish aphid that does not readily drop from disturbed foliage. All stages of the aphid, except for the egg stage, can transmit RLCV. The alpha strain is acquired after a minimum of 2 hr and retained by the aphid for life. Single-aphid transmission efficiency is 19%. Minimum acquisition access time, maximum retention time, and transmission efficiency of single aphids for the beta strain of RLCV have not been determined. Inoculation threshold times in *A. rubicola* have not been determined for RLCV.

In studies of *A. rubicola* in New York, aphid populations reached a first peak in late July and a second, larger peak in early October. Virus spread in the field was primarily in the direction of prevailing winds and greatest near sources of viruliferous aphids. The small European aphid, *A. idaei* van der Goot, is an experimental vector of RLCV. It occurs in coastal British Columbia but has not been reported elsewhere in North America. Other aphids experimentally tested have failed to transmit the virus.

Control

RLCV control involves the use of certified stock planted as far away as possible from wild *Rubus* (especially blackberries that may be asymptomatic) or from cultivated *Rubus* infected with RLCV. Although ineffective in preventing primary RLCV infection by *A. rubicola*, approved aphicide sprays aid in reducing established populations of this aphid in *Rubus* fields, thus slowing the rate of secondary spread by wingless aphids.

Because RLCV usually spreads slowly, the annual roguing of plants showing leaf curl symptoms is an effective control procedure. Unless the rate of infection is unusually slow, it is usually unprofitable to keep a planting after more than 10% of it has been rogued out.

Some red and black raspberry breeding lines are known to be immune or highly resistant to RLCV or to its vector aphid. It would be possible to breed for immunity or strong resistance to this disease in commercial cultivars, but such breeding has not yet been done. The purple raspberry cultivar Royalty, released by the New York Agricultural Experiment Station, is resistant to colonization by *A. rubicola*.

Selected References

Bennett, C. W. 1930. Further observations and experiments on the curl disease of raspberries. Phytopathology 20:787-802.
Bolton, A. T. 1970. Spread of raspberry leaf curl virus. Can. J. Plant Sci. 50:667-671.
Converse, R. H. 1962. Insect and graft transmissions of alpha- and beta-curl viruses of raspberries. (Abstr.) Phytopathology 52:728.
Sanford, J. C., and Ourecky, D. K. 1983. 'Royalty' purple raspberry. HortScience 18:109-110.
Schaefers, G. A. 1967. Aphid control in virus-free raspberry nursery stock. Farm Res. 33:14-15.
Stace-Smith, R. 1962. Studies on *Rubus* virus diseases in British Columbia. 8. Raspberry leaf curl. Can. J. Bot. 40:651-657.
Stace-Smith, R., and Converse, R. H. 1987. Raspberry leaf curl. Pages 189-190 in: Virus Diseases of Small Fruits. R. H. Converse, ed. U.S. Dep. Agric. Agric. Handb. 631.

(Prepared by R. H. Converse)

Cucumber Mosaic

Cucumber mosaic virus (CMV) has been reported on red raspberry in Scotland and the Soviet Union. Its occurrence in *Rubus* is rare and seems to cause little damage. Except for one early, unconfirmed report, CMV has not been reported on *Rubus* in North America. This virus has one of the widest host ranges and is one of the most widespread of all plant viruses, infecting many crops and wild plants worldwide.

Symptoms

On red raspberry in Scotland, CMV causes symptoms that vary with cultivar and location. Some plants may be symptomless, whereas others may have symptoms ranging from pale green blotches to chlorotic blotches and line patterns on leaves (Plate 59) to the development of small leaves with bright, chlorotic mottling. A green mosaic develops on black raspberry (*R. occidentalis* L.) graft-inoculated with CMV. The virus is symptomless in cultivated blackberry, but it caused severe leaf distortion, chlorotic blotching, decline in vigor, and plant death in naturally infected wineberry (*R. phoenicolasius* Maxim.) in Scotland.

CMV isolates from *Rubus* were readily transmitted in Scotland by sap inoculation in 2% nicotine alkaloid solution to a number of common herbaceous greenhouse virus test plants.

Causal Agent

CMV is the type member of the cucumovirus group. It is nonpersistently transmitted by many aphid species. It has isometric virus particles (approximately 28 nm in diameter), with a single coat protein subunit with a relative molecular mass of 24,500 Da. The particles contain four single-stranded RNA (ssRNA) species (five if a satellite RNA is also present).

CMV is a relatively unstable virus, but monoclonal and polyclonal antisera have been used against many of the known serotypes. All of the *Rubus* isolates of CMV tested in Scotland were serologically indistinguishable from the W strain of CMV, which is common in many crop plants in the United Kingdom. CMV can be readily detected by enzyme-linked immunosorbent assay (ELISA) directly from infected *Rubus* sap, provided a homologous antiserum is selected.

Epidemiology and Control

Although CMV is readily transmitted in a nonpersistent manner by many aphid species to many plant species, only *Aphis idaei* van der Goot has been shown to transmit the virus to red raspberry. CMV is seedborne in some plant species, but not in wineberry. In greenhouse studies in Australia with peppers, the virus was experimentally transmitted in the absence of vectors from CMV-infected plant debris in the soil.

In *Rubus*, the virus has occurred too rarely to warrant the development of control measures. Mother plants for *Rubus* certification programs should be tested for the presence of latent CMV infection by sap inoculation to *Chenopodium quinoa* Willd. and *Nicotiana clevelandii* A. Gray, with follow-up serological identification by ELISA.

Selected References

Francki, R. I. B., Mossop, D. W., and Hatta, T. 1979. Cucumber mosaic virus. Descriptions of Plant Viruses, No. 213. Commonwealth Mycological Institute and Association of Applied Biologists, Kew, Surrey, England. 6 pp.

Harrison, B. D. 1958. Cucumber mosaic virus in raspberry. Plant Pathol. 7:109-111.

Jones, A. T. 1976. An isolate of cucumber mosaic virus from *Rubus phoenicolasius* Maxim. Plant Pathol. 25:137-140.

Jones, A. T. 1987. Cucumber mosaic virus in raspberry. Pages 191-192 in: Virus Diseases of Small Fruits. R. H. Converse, ed. U.S. Dep. Agric. Agric. Handb. 631.

Kennedy, J. S., Day, M. F., and Eastop, V. F. 1962. A Conspectus of Aphids as Vectors of Plant Viruses. Commonwealth Institute of Entomology, London. 114 pp.

Pares, R. D., and Gunn, L. V. 1989. The role of non-vectored soil transmission as a primary source of infection by pepper mottle and cucumber mosaic viruses in glasshouse-grown *Capsicum* in Australia. J. Phytopathol. 126:353-360.

Tomlinson, J. A., Carter, A. L., Faithfull, E. M, and Webb, M. J. W. 1973. Purification and serology of the W strain of cucumber mosaic virus. Ann. Appl. Biol. 74:181-189.

(Prepared by R. H. Converse)

Raspberry Vein Chlorosis

Raspberry vein chlorosis virus (RVCV) is widespread in Europe and the Soviet Union and has been reported in New Zealand and infrequently in British Columbia. In Europe it is of commercial importance on red raspberry, causing fewer, shorter canes and early ripening. Plant vigor is further depressed when RVCV occurs in combination with other viruses in raspberry. RVCV has caused pollen abortion, has retarded embryo sac development, and has decreased berry weight in red raspberry experimentally infected in British Columbia.

Symptoms

In field-infected red raspberry, RVCV causes chlorosis of minor leaf veins (Plate 60). Epinasty and leaf distortion may also occur. Natural infections of blackberry and blackberry-raspberry hybrids are not known; symptoms of RVCV experimentally induced on loganberry were similar to those on red raspberry cultivars. In Scotland, blackberry and black raspberry cultivars were immune from graft-inoculation with RVCV. *Fragaria vesca* L. var. *semperflorens* (Duch.) Ser. 'Alpine' can be graft-inoculated and may develop local minor vein necrosis as well as vein chlorosis symptoms, such as those in red raspberry.

Causal Agent

RVCV has been shown by electron microscopy to be associated with large bacilliform particles (430–560 × 65–91 nm), rounded at either one or both ends, which occur in the cytoplasm of vascular parenchyma cells in plant hosts and in the cytoplasm of aphid brain, salivary gland, and other cells. This virus has not been successfully mechanically inoculated and is probably a rhabdovirus.

Epidemiology and Control

RVCV is transmitted by the small European raspberry aphid, *Aphis idaei* van der Goot, but not by the aphid vectors of the raspberry mosaic virus complex, *Amphorophora agathonica* Hottes and *Amphorophora idaei* Börner. RVCV is probably persistent in its aphid vector. Maximum transmission efficiency was 46% after an acquisition access period of 7 days and an inoculation access period of 30 days.

In Europe, old red raspberry cultivars are commonly infected with RVCV. This virus is not readily eliminated from infected plants by experimental thermotherapy, but clones free of RVCV have been developed by following thermotherapy with shoot apex tissue culture and plant regeneration. Certain immune red and black raspberry cultivars provide the basis of research in Europe aimed at developing RVCV-immune commercial cultivars.

Selected References

Baumann, G. 1982. Elimination of a heat-stable raspberry virus by combining heat treatment and meristem culture. Acta Hortic. 129:11-12.

Jennings, D. L., and Jones, A. T. 1986. Immunity from raspberry vein chlorosis virus in raspberry and its potential for control of the virus through plant breeding. Ann. Appl. Biol. 108:417-422.

Jones, A. T. 1980. Some effects of latent virus infection in red raspberry. Acta Hortic. 95:63-70.

Jones, A. T., Murant, A. F., and Stace-Smith, R. 1977. Raspberry vein chlorosis virus. Descriptions of Plant Viruses, No. 174. Commonwealth Mycological Institute and Association of Applied Biologists, Kew, Surrey, England. 4 pp.

Murant, A. F., and Roberts, I. M. 1980. Particles of vein chlorosis virus in the aphid vector, *Aphis idaei*. Acta Hortic. 95:31-35.

(Prepared by R. H. Converse)

Leafhopper-Transmitted Diseases

Rubus Stunt

Rubus stunt is a very severe disease that naturally affects only plants in the genus *Rubus*; no immune *Rubus* germ plasm has been reported. *Apium*, *Chrysanthemum*, *Fragaria*, and *Trifolium* have been used as experimental hosts. The disease occurs in wild and cultivated *Rubus* in Europe, the Soviet Union, and the Middle East. Crop-threatening epidemics have occurred in some areas of Europe and the Soviet Union. Rubus stunt is not known to occur in North America, but *Macropsis fuscula* Zett., the major leafhopper vector of Rubus stunt in Europe, has become well established in western North America in wild blackberry, especially in *R. procerus* P. J. Müll.

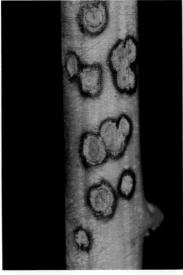

1. Anthracnose lesions on black raspberry primocane. (Courtesy B. Williamson; copyright Scottish Crop Research Institute)

2. Deeply pitted anthracnose lesions on floricane of red raspberry cv. Glen Clova. (Courtesy B. Williamson; copyright Scottish Crop Research Institute)

3. Anthracnose lesions on tayberry leaves. (Courtesy B. Williamson; copyright Scottish Crop Research Institute)

4. Red raspberry fruit affected by anthracnose. (Courtesy B. Williamson; copyright Scottish Crop Research Institute)

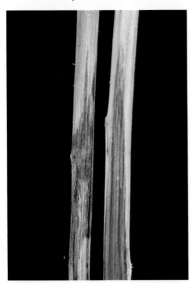

5. Brown, spreading, vascular lesions caused by *Leptosphaeria coniothyrium* (cane blight) on red raspberry floricanes, visible after scraping off bark. (Courtesy B. Williamson; copyright Scottish Crop Research Institute)

6. Cane blight lesion on red raspberry floricane. Note wound (center) and sooty black spore deposits on cane surface. (Courtesy B. Williamson; copyright Scottish Crop Research Institute)

7. Spur blight symptoms on red raspberry leaflets from primocanes. (Courtesy B. Williamson; copyright Scottish Crop Research Institute)

COLOR PLATES

8. Spur blight lesions on red raspberry primocanes. (Courtesy B. Williamson; copyright Scottish Crop Research Institute)

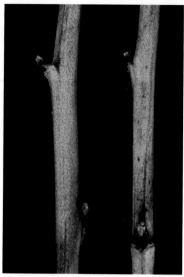

9. Spur blight lesions on overwintered red raspberry floricanes. Note silvered lesions covered by black pseudothecia. (Courtesy B. Williamson; copyright Scottish Crop Research Institute)

10. *Botrytis cinerea* attacking young primocane of red raspberry cv. Chilliwack. (Courtesy P. R. Bristow)

11. Lesions on red raspberry primocanes caused by *Botrytis cinerea*. (Courtesy B. Williamson; copyright Scottish Crop Research Institute)

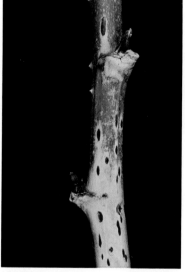

12. Black sclerotia on white lesion caused by *Botrytis cinerea* on red raspberry floricane. (Courtesy B. Williamson; copyright Scottish Crop Research Institute)

13. Blackberry cane in spring affected by purple blotch (*Septocyta ruborum*). (Courtesy B. Williamson; copyright Scottish Crop Research Institute)

14. Blackberry cane in spring affected by Ascospora dieback (*Clethridium corticola*). (Courtesy B. Williamson; copyright Scottish Crop Research Institute)

15. Cane canker on thornless blackberry, caused by *Botryosphaeria dothidea*.

16. Blackberry cane severely affected by rosette, or double blossom (caused by *Cercosporella rubi*), and showing witches'-broom. (Courtesy R. K. Jones)

17. An early proliferation of leafy shoots from rosette-affected bud infected by *Cercosporella rubi* during previous growing season. (Courtesy R. K. Jones)

18. Distortion of blackberry flowers affected by rosette, or double blossom, caused by *Cercosporella rubi*. (Courtesy R. K. Jones)

19. Boysenberry leaf affected by downy mildew (*Peronospora sparsa*). Note angular lesions spreading along veins.

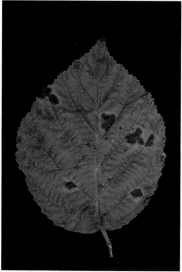

20. Wild blackberry leaf with localized lesions caused by *Peronospora sparsa*. (Courtesy B. Williamson; copyright Scottish Crop Research Institute)

21. Downy mildew (dryberry) symptoms on boysenberry fruit.

22. Powdery mildew on blackberry, caused by *Sphaerotheca macularis*.

23. Blackberry leaves severely affected by powdery mildew. Note distortion on infected leaves.

24. Raspberry leaf spot, caused by *Sphaerulina rubi*.

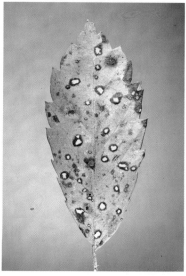

25. Septoria leaf spot on leaflet of *Rubus trivialis*.

26. Gray mold, caused by *Botrytis cinerea*, on red raspberry fruit. Note dusty covering of gray spores on infected fruit.

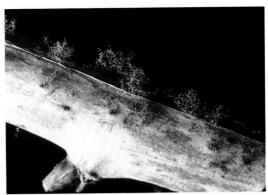

27. Production of conidia of *Botrytis cinerea* from sclerotia on a red raspberry floricane. (Courtesy B. Williamson; copyright Scottish Crop Research Institute)

28. Stamen blight symptoms on infected red raspberry flower (left) next to healthy flower (right). Fungus is sporulating on infected flower parts.

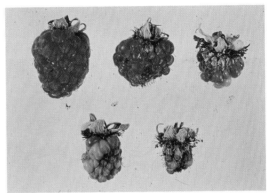

29. Malformed red raspberry fruit affected with stamen blight.

30. Early-season symptoms of orange rust. Note proliferation of weak, spindly shoots from infected black raspberry floricane in early spring. (Courtesy M. A. Ellis)

31. Blisterlike aecia covering lower surface of black raspberry leaf affected by orange rust. (Courtesy L. V. Madden and S. Heady)

32. Cane and leaf rust pustules (uredinia) on blackberry floricane.

COLOR PLATES

33. Cane and leaf rust pustules (uredinia) on lower surface of blackberry leaf infected by *Kuehneola uredinis*.

34. Yellow pustules (aecia) on upper surface of red raspberry leaf infected by *Phragmidium rubi-idaei* (yellow rust). (Courtesy B. Williamson; copyright Scottish Crop Research Institute)

35. Close-up of yellow rust aecia on red raspberry leaf. (Courtesy B. Williamson; copyright Scottish Crop Research Institute)

36. Yellow rust pustules (uredinia) on lower leaf surface of red raspberry. (Courtesy B. Williamson; copyright Scottish Crop Research Institute)

37. Symptoms of late leaf rust on lower surface of red raspberry leaf infected by *Pucciniastrum americanum*. Note yellow pustules (uredinia).

38. Yellow pustules (uredinia) on a red raspberry infected by *Pucciniastrum americanum*.

39. Aecia of *Pucciniastrum americanum* on white spruce (*Picea glauca*).

40. Leaf spots on upper surface of blackberry leaf infected by *Phragmidium violaceum* (blackberry rust). (Courtesy B. Williamson; copyright Scottish Crop Research Institute)

41. Yellow pustules (uredinia) on lower surface of blackberry leaf infected by *Phragmidium violaceum* (blackberry rust). (Courtesy B. Williamson; copyright Scottish Crop Research Institute)

42. Aerial photograph of red raspberry planting in Scotland. Large areas of missing plants are caused by Phytophthora root rot. (Courtesy J. Duncan; copyright Scottish Crop Research Institute)

43. Aboveground symptoms of Phytophthora root rot on red raspberry primocane.

44. Aboveground symptoms of Phytophthora root rot on floricanes of red raspberry. (Courtesy B. Williamson; copyright Scottish Crop Research Institute)

45. Belowground symptoms of Phytophthora root rot on crown and roots of red raspberry plant. Note sharp line of demarcation between healthy (white) and infected (reddish brown) tissues. (Courtesy M. A. Ellis)

46. Black raspberry plant severely affected by Verticillium wilt.

47. Red raspberry canes killed by *Armillaria* sp. spreading from a nearby hedgerow. Note how disease is spreading down row. (Courtesy B. Williamson; copyright Scottish Crop Research Institute)

48. Diagnostic white mycelium of *Armillaria* sp. exposed beneath dead bark at crown of affected red raspberry plant. (Courtesy B. Williamson; copyright Scottish Crop Research Institute)

49. White root rot of red raspberry, caused by *Vararia* spp.

50. Young, fleshy galls caused by *Agrobacterium tumefaciens* on black raspberry cane affected by crown gall. (Courtesy M. A. Ellis)

51. Hard, woody galls caused by *Agrobacterium tumefaciens* near crown of old black raspberry floricane. (Courtesy M. A. Ellis)

52. Fire blight symptoms on tip of raspberry floricane infected with *Erwinia amylovora*. Note shepherd's crook symptom at tip of cane. (Courtesy S. N. Jeffers)

53. Fire blight symptoms on blackberry fruit, caused by *Erwinia amylovora*. Note brown, dried fruit.

54. Symptoms of Pseudomonas blight on red raspberry floricane. (Photo by H. S. Pepin; copyright Department of Agriculture, Government of Canada)

55. Mottling on leaflet of Munger black raspberry seedling infected with raspberry mosaic disease.

56. Black raspberry leaf showing fine veinclearing on lower leaflet (right) and right side of terminal leaflet caused by Rubus yellow net virus. (Reprinted, by permission, from Stace-Smith and Jones, 1987)

57. Symptoms of dwarfing, chlorosis, and leaf distortion in Lloyd George red raspberry chronically infected with raspberry leaf curl virus. (Reprinted, by permission, from Stace-Smith and Converse, 1987)

58. Symptoms of leaf chlorosis and distortion in New Logan black raspberry chronically infected with raspberry leaf curl virus. (Reprinted, by permission, from Stace-Smith and Converse, 1987)

COLOR PLATES

59. Chlorotic blotches caused by cucumber mosaic virus in unnamed red raspberry selection in Scotland. (Copyright Scottish Crop Research Institute)

60. Chlorosis of minor veins of Glen Prosen red raspberry leaf infected with raspberry vein chlorosis virus. (Reprinted, by permission, from Jones et al, 1977; copyright Scottish Crop Research Institute)

61. *Rubus laciniatus* 'Thornless Evergreen' floricane affected by Rubus stunt disease, showing witches'-broom and chlorosis. (Reprinted, by permission, from van der Meer, 1987)

62. Mottling symptoms on leaf of Glen Clova red raspberry infected with Arabis mosaic virus. (Reprinted, by permission, from Murant, 1981b; copyright Scottish Crop Research Institute)

63. Delayed foliation of Puyallup red raspberry chronically infected with tomato ringspot virus (left) compared with uninfected plant (right). (Reprinted, by permission, from Stace-Smith and Converse, 1987)

64. Red raspberry plant affected with tomato ringspot virus, showing foliar ring spots and line patterns. (Courtesy F. D. McElroy)

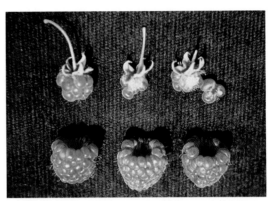

65. Crumbly fruit (top) of red raspberry infected with tomato ringspot virus compared to healthy fruit (bottom). (Courtesy F. D. McElroy)

66. Local lesions on leaf of *Chenopodium quinoa* mechanically inoculated with tomato ringspot virus. (Reprinted, by permission, from Stace-Smith and Converse, 1987)

67. Systemic mottling on cucumber mechanically inoculated with tomato ringspot virus.

68. Necrotic ringspot lesions on leaves of *Nicotiana tabacum* 'Haranova' mechanically inoculated with tomato ringspot virus. (Reprinted, by permission, from Stace-Smith and Converse, 1987)

69. Ring spots and line patterns on a leaf of Meeker red raspberry infected with raspberry bushy dwarf virus. (Courtesy H. A. Daubeny, Agriculture Canada)

COLOR PLATES

70. Ring and line patterns on *Chenopodium quinoa* leaves mechanically inoculated with raspberry bushy dwarf virus.

71. Brilliant yellow blotches and ring patterns in leaf of Schoenemann red raspberry naturally infected with apple mosaic virus. (Reprinted, by permission, from Baumann et al, 1987)

72. Leaf of Thornless Loganberry infected with blackberry calico virus.

73. Leaves of Marcy red raspberry naturally infected in New Zealand with cherry leaf roll virus, showing distortion, yellow blotches, and ring spots. (Courtesy G. A. Wood; copyright New Zealand Department of Scientific and Industrial Research)

74. Necrotic rings on leaves of *Nicotiana tabacum* 'Haranova' mechanically inoculated with *Rubus* strain of tobacco streak virus. (Reprinted, by permission, from Stace-Smith, 1987)

75. Necrotic local lesions in leaf of *Chenopodium amaranticolor* 20 days after mechanical inoculation with wineberry latent virus. (Reprinted, by permission, from Jones, 1987; copyright Scottish Crop Research Institute)

76. Mild mottling associated with Alpine mosaic agent on *Fragaria vesca* var. *semperflorens* 'Alpine' after petiole insert leaflet grafting from Darrow blackberry.

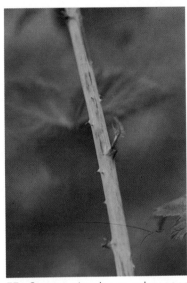

77. Severe streak on primocane of wild black raspberry *Rubus occidentalis*.

78. Boysenberry cane affected with boysenberry decline disease (right), showing compact rosettes of proliferated lateral shoots with undersized leaves; normal boysenberry cane (left). (Courtesy G. A. Wood; copyright New Zealand Department of Scientific and Industrial Research)

79. Boysenberry flower affected with boysenberry decline disease (right), showing crinkled petals, stubby stamens, and enlarged styles and stigmas; normal boysenberry flower (left). Some petals were removed from both flowers for photography. (Courtesy G. A. Wood; copyright New Zealand Department of Scientific and Industrial Research)

80. Boysenberry fruit affected with boysenberry decline disease (right), showing abnormal cone shape and persistent styles that give fruit a hairy appearance; immature normal boysenberry fruit (left). (Courtesy G. A. Wood; copyright New Zealand Department of Scientific and Industrial Research)

81. Chlorotic ring spots and line patterns on leaves of *Rubus parviflorus* (thimbleberry) infected with thimbleberry ringspot virus. (Reprinted, by permission, from Stace-Smith, 1987a)

82. Mottling and leaf distortion in wild blackberry, caused by tobacco ringspot virus. (Courtesy G. V. Gooding, Jr.; reprinted from Stace-Smith, 1987b)

83. Three-year-old Meeker raspberry plants in various stages of decline due to feeding of *Pratylenchus penetrans*. Note dead stems on center plant caused by increased winter injury from nematode attack.

84. Three-year-old plantation of Willamette red raspberry in advanced stages of decline due to high populations of *Pratylenchus penetrans* at time of planting. Note declining plants at margin and areas where plants never established.

85. Damage to raspberry roots caused by *Xiphinema bakeri*. Note characteristic terminal swellings and "fishhook" symptoms.

86. Adult female of the raspberry crown borer (*Pennisetia marginata*).

87. Larva of the raspberry crown borer.

88. Adult black vine weevil (*Otiorhynchus sulcatus*). (Courtesy D. G. Nielsen)

89. Clay-colored weevil (*Otiorhynchus singularis*) of the United Kingdom, feeding on red raspberry leaf petiole. (Courtesy S. C. Gordon; copyright Scottish Crop Research Institute)

90. Female strawberry crown moth (*Synanthedon bibionipennis*).

91. Japanese beetles (*Popillia japonica*) attacking ripe fruit of red raspberry. (Courtesy T. L. Ladd, Jr.)

92. Adult green June beetle (*Cotinis nitida*).

COLOR PLATES

93. Adult rose chafer (*Macrodoctylus subspinosus*).

94. Adult tarnished plant bug (*Lygus lineolaris*) on ripe red raspberry fruit.

95. Tarnished plant bug injury to raspberry fruit.

96. Adult raspberry bud moth (*Lampronia rubiella*). (Courtesy S. C. Gordon; copyright Scottish Crop Research Institute)

97. Mature raspberry bud moth larva feeding in red raspberry bud in the spring. (Note frass.) (Courtesy S. C. Gordon; copyright Scottish Crop Research Institute)

98. Adult picnic beetle (*Glischrochilus quadrisignatus*).

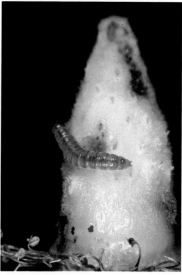

99. Larva of raspberry fruitworm on a raspberry receptacle after harvest. (Courtesy S. C. Gordon; copyright Scottish Crop Research Institute)

COLOR PLATES

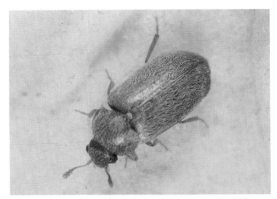

100. Adult raspberry fruitworm (*Byturus unicolor*).

101. Yellowjackets attacking ripe red raspberry fruit.

102. Damage to red raspberry flower bud (center), caused by strawberry bud weevil, or clipper. Note that pedicel has been severed. (Courtesy A. M. Agnello, New York State Agricultural Experiment Station)

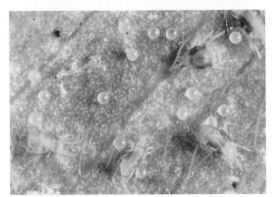

103. Adult spider mites and eggs on undersurface of raspberry leaf.

104. Normal red raspberry leaf (right); spider mite-infested leaf (left). (Courtesy R. N. Williams)

105. Partial colony of overwintering dryberry mites in red raspberry bud. Outer bud scales have been removed. (Courtesy S. C. Gordon; copyright Scottish Crop Research Institute)

106. Effect of dryberry mite feeding on leaves of tayberry (left), as compared to normal foliage (right). (Courtesy S. C. Gordon; reprinted, by permission, from Journal of Horticultural Science, 59, 525 [1984])

107. Large raspberry aphid (*Amphorophora agathonica*) adult (center) and nymphs.

108. Small raspberry aphid (*Aphis rubicola*) adult (center) and nymph.

109. Adult of orange tortrix (*Argyrotaenia citrana*), a leaf roller.

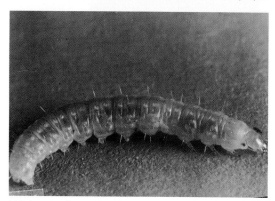

110. Larva of orange tortrix.

111. Red-banded leaf roller (*Argyrotaenia velutinana*) damage on red raspberry leaf. (Courtesy R. N. Williams)

112. Climbing cutworm damage to red raspberry.

113. Lateral view of adult blackberry psyllids.

114. Blackberry psyllid damage to black raspberry foliage. (Courtesy R. N. Williams)

115. Adult rednecked cane borer (*Agrilus ruficollis*). (Courtesy R. N. Williams)

116. Rednecked cane borer damage on wild blackberry cane. (Courtesy R. N. Williams)

117. Larvae of raspberry cane midge (*Resseliella theobaldi*) feeding on periderm of red raspberry. Bark has been removed to expose larvae. (Courtesy S. C. Gordon; copyright Scottish Crop Research Institute)

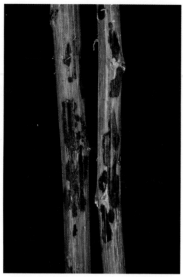

118. Patch lesions on primocanes of red raspberry, indicating raspberry cane midge larvae feeding sites and extent of fungal colonization (midge blight). (Courtesy S. C. Gordon; copyright Scottish Crop Research Institute)

119. Close-up of raspberry cane borer damage on red raspberry primocane.

120. Oviposition punctures of tree cricket on blackberry cane (top) and lateral view of exposed eggs (bottom).

121. Rose scale (*Aulacaspis rosae*) on lower portion of *Rubus* cane.

122. Interveinal yellowing in Heritage raspberry caused by high rate of simazine applied to planting in early spring. (Courtesy R. C. Funt)

123. Leaf yellowing and interveinal necrosis of floricane leaves caused by excessive terbacil application.

124. Interveinal necrosis on primocane leaves caused by excessive terbacil application.

125. Bleached veinal areas of older leaves, typical of norflurazon injury.

126. Glyphosate applied in spring or summer is transported to growing points of *Rubus* plants, causing yellowing of terminal leaves.

127. Later stages of glyphosate injury, characterized by death of growing points.

128. Straplike and puckered leaves caused by glyphosate applied in previous years. (Courtesy J. L. Maas)

129. Paraquat injury, characterized by rapid browning of leaves at areas of spray contact. Symptoms developed within 48 hr of application.

130. Abnormal plant growth caused by glufosinate applied in early summer to primocanes.

131. Upward curling of leaves and meristems caused by drift from 2,4-D.

132. Upward curling of leaves and leaf necrosis caused by dicamba. Symptoms developed within 48 hr of application.

133. Winter injury on red raspberry. Note dead floricanes in contrast to emergence of healthy primocanes in spring. (Courtesy B. Williamson; copyright Scottish Crop Research Institute)

134. Winter injury on a lateral bud of Meeker red raspberry. Bud tissues are brown and dead. (Courtesy B. C. Strik)

135. Cross section of Meeker red raspberry floricane, showing vascular damage from winter injury. Note healthy green tissue compared to brown, discolored tissues. (Courtesy B. C. Strik)

136. Sunburn, or sunscald, on thornless blackberry (cv. Hull Thornless).

137. White drupelet disorder of red raspberry, caused by combination of exposure to high temperature and ultraviolet radiation. (Courtesy M. A. Ellis)

(Himalaya blackberry). At present, importation of vegetative *Rubus* plant material into the United States from Europe is restricted, partly because of the threat posed to the U.S. *Rubus* industry by this disease. Although the disease occurs only in Europe, it poses a threat to North American *Rubus* crops.

Symptoms

All infected *Rubus* plants produce numerous spindly shoots (witches'-broom) (Plate 61) and flower phyllody and proliferation (Fig. 24). No cultivars are known in which chronic Rubus stunt disease is fully latent, but infected plants usually remain symptomless until a year after infection. Plants of the British red raspberry cultivar Malling Promise chronically infected with Rubus stunt rarely exhibit flower proliferation symptoms. The British red raspberry cultivar Malling Landmark and the U.S. blackberry (*R. laciniatus* Willd.) cultivar Thornless Evergreen are very sensitive indicators for verification of stunt infection by graft analysis but require a year to develop symptoms after grafting. Plant mycoplasmalike organisms (MLOs) are generally unevenly distributed in infected host phloem tissues. The titer of many plant MLOs is higher and more seasonally constant in roots than in shoots; therefore, it seems reasonable to include root samples in tests for Rubus stunt.

Causal Agent

This disease is caused by an MLO. The pathogen is phloem-limited (Fig. 25) and has not been cultured; its serological and biochemical relationships to other plant MLOs are unknown.

A lethal disease causing witches'-broom and green petal symptoms was discovered in 1981 in one commercial black raspberry field of the cultivar Munger in Oregon. Typical MLO bodies were seen by thin-section electron microscopy in phloem sieve tubes of infected plants from that field. The disease did not infect red raspberry plants adjoining the infected black raspberry field, despite the presence of abundant *M. fuscula* leafhoppers. The relationship between Munger witches'-broom and European Rubus stunt is unknown.

Epidemiology

In Europe, Rubus stunt disease is limited to wild and cultivated *Rubus* spp. Its major vectors, *M. fuscula* and some other *Macropsis* spp., are limited to *Rubus*, on which they complete one generation per year. *M. fuscula* overwinters in the egg stage, but Rubus stunt is not transovarially transmitted. Larvae may acquire the Rubus stunt MLO but cannot transmit it until an 8-week latent period has elapsed. Once infected, *M. fuscula* remains capable of transmitting the MLO for life.

In addition to *Macropsis* spp., a froghopper (*Philaenus spumarius* L.) and two leafhoppers (*Allygus mayri* Kbm. and *Euscelis plebeja* Fallén) have been reported in Hungary to be experimental vectors. The latter three insects do not specifically live on *Rubus* and are probably not important in the natural spread of Rubus stunt.

Control

Rubus stunt disease is controlled by the use of certified planting stock. In Europe it is also controlled by insecticides, which destroy overwintering *M. fuscula* eggs and kill various instars of this leafhopper during summer, when acquired Rubus stunt MLOs are still latent in the insect. These leafhoppers become able to transmit the MLO in August and September. This MLO can be eliminated from vegetative *Rubus* by treatment in hot water (45°C) for 2–3 hr.

Selected References

Converse, R. H., Clarke, R. G., Oman, P. W.,Sr., and Milbrath, G. M. 1982. Witches'-broom disease of black raspberry in Oregon. Plant Dis. Rep. 66:949-951.

Murant, A. F., and Roberts, I. M. 1971. Mycoplasma-like bodies associated with Rubus stunt disease. Ann. Appl. Biol. 67:389-393.

Seemüller, E., Schaper, U., and Zimbelmann, F. 1984. Seasonal variations in colonization patterns of mycoplasmalike organisms associated with apple proliferation and pear decline. Z. Pflanzenkrankh. Pflanzenschutz 91:371-382.

van der Meer, F. A. 1987. Rubus stunt. Pages 197-203 in: Virus Diseases of Small Fruits. R. H. Converse, ed. U.S. Dep. Agric. Agric. Handb. 631.

(Prepared by R. H. Converse)

Nematode-Transmitted Diseases

European Nepovirus Diseases

Four isometric viruses that are vectored by nematodes and belong to the nepovirus group occur in commercial red raspberry fields in Europe and are considered to cause major losses of stand and yield when present. These four viruses can be divided into two groups with different nematode vectors:

Fig. 24. Phyllody of wild blackberry flower affected by Rubus stunt disease. (Reprinted, by permission, from van der Meer, 1987)

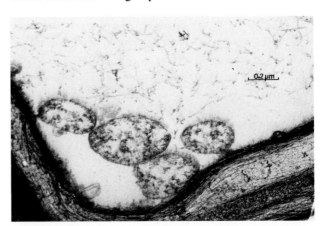

Fig. 25. Mycoplasmalike organisms associated with Rubus stunt disease in sieve tube of diseased red raspberry. (Reprinted, by permission, from Murant and Roberts, 1971; copyright Scottish Crop Research Institute)

1) *Xiphinema*-vectored Arabis mosaic virus (AMV) and strawberry latent ringspot virus (SLRV) and 2) *Longidorus*-vectored raspberry ringspot virus (RRV) and tomato black ring virus (TBRV). None of these viruses are known to occur naturally in North America, although some have been introduced in grapevines and other crops. The federal plant quarantines in effect in Canada and the United States governing importation of *Rubus* plants from Europe are designed, in part, to prevent these viruses from becoming established in North America. All four have wide host ranges in many flowering plant families in annual and perennial plants and in their seeds, including crop and weed species. Many *Rubus* and strawberry cultivars are susceptible to the viruses.

Symptoms

On red raspberry cultivars in the United Kingdom, AMV and SLRV cause various types of leaf symptoms (depending on the cultivar, the virus, and the virus strain), which include yellow speckling, mottling, vein yellowing, and down-curling (Plate 62). Both viruses cause a disease of red raspberry called raspberry yellow dwarf. Infected red raspberry plants are stunted and produce little or no fruit. In areas where these viruses are spread in the soil by viruliferous nematodes, roughly circular, expanding areas of weak plants are seen in infected red raspberry fields. Leaves of the blackberry (*R. procerus* P. J. Müll.) cultivar Himalaya Giant in Scotland infected with AMV develop a yellow mosaic. AMV and SLRV are rare in Scotland but cause severe disease in England.

RRV and TBRV cause a disease of red raspberry in Scotland called raspberry leaf curl, which is completely different from the aphid-borne virus disease of raspberries in North America known by the same name. Except during hot summer weather, RRV may cause conspicuous chlorotic ring spots on leaves of infected red raspberry cultivars (Fig. 26). In some cultivars, it causes severe down-curling of brittle leaves (Fig. 27). Severe dwarfing and plant death occur in some infected cultivars, and a depression of vigor in infected but symptomless plants occurs in others. TBRV causes relatively mild symptoms in most susceptible red raspberry cultivars, but some, such as Malling Exploit, develop faint foliar chlorotic rings and spots when first infected (Fig. 28) and become progressively more dwarfed and produce crumbly fruit in subsequent years.

When mechanically inoculated with sap from infected *Rubus* leaves, all four European nepoviruses readily infect several standard herbaceous greenhouse test plants, including *Chenopodium amaranticolor* Coste & Reyn. and *Nicotiana tabacum* L. 'White Burley.' These and many other herbaceous indicator plants develop nondiagnostic symptoms after inoculation with the viruses.

Causal Agents

AMV, RRV, and TBRV are members of the nepovirus group. SLRV, although also nematode-transmitted, is a tentative member, because it differs from the other three viruses in its coat protein subunits. All four viruses are readily sap-transmitted to common greenhouse herbaceous test plants and are relatively easy to purify or use for serological tests from selected herbaceous hosts. All four viruses are isometric, 28–30 nm in diameter, and each has two species of single-stranded RNA (ssRNA), with relative molecular masses (M_r) ranging from $2.9–2.7 \times 10^6$ Da for the largest of the viruses to $1.7–1.3 \times 10^6$ Da for the smallest. For RRV and TBRV, the coat protein gene is known to be carried in the smaller ssRNA of the bipartite genome. Some isolates of AMV, SLRV, and TBRV each have a satellite ssRNA of M_r 500,000 Da. The nucleotide sequences for all four viruses have been determined, and a cDNA probe has been developed for AMV.

Polyclonal and monoclonal antisera against these viruses have been developed, and all of the viruses can be readily detected in sap from infected *Rubus* leaves by standard serological tests, including ELISA. These viruses are serologically unrelated, and serologically recognizable strain differences have been found in each.

AMV, SLRV, RRV, and TBRV are most easily identified in suspect *Rubus* plants by serological means (usually ELISA) or by use of cDNA probes.

Epidemiology

AMV and SLRV are vectored by the dagger nematode *Xiphinema diversicaudatum* (Micoletzky) Thorne, whose

Fig. 26. Chlorotic ringspot symptoms in leaf of Malling Jewel red raspberry infected with raspberry ringspot virus. (Reprinted, by permission, from Cadman, 1956; copyright Scottish Crop Research Institute)

Fig. 27. Leaf curl symptoms in Norfolk Giant red raspberry infected with raspberry ringspot virus. (Reprinted, by permission, from Murant, 1981; copyright Scottish Crop Research Institute)

larvae and adults can acquire either virus within 1 day, transmit them within 3 days, and retain them in fallow soil for 3 months or more. This nematode does not retain either virus through moulting or transmit them through its eggs. In the nematode, both viruses appear to be very specifically attached as monolayers to the cuticular lining of the basal portion of the stylet, esophagus, and esophageal bulb. Neither virus appears to multiply in the nematode or occur in it, except in its alimentary tract. Strains of this nematode that are efficient vectors of local isolates but poor vectors of Scottish isolates of SLRV exist in southern Europe.

RRV and TBRV have the needle nematode, *Longidorus*, as a natural vector, and three species are involved in transmission. *L. elongatus* de Man transmits local strains of both RRV and TBRV in Scotland. In England and continental Europe, *L. macrosoma* Hooper transmits a strain of RRV that is serologically distinct from the one in Scotland, and *L. attenuatus* Hooper transmits a serologically distinct strain of TBRV in those areas. These needle nematodes can retain RRV and TBRV for as long as 9 weeks, but they lose the viruses when molting and do not transmit them in their eggs. The viruses appear to be associated in a very specific manner with the stylet-guiding sheath but probably do not multiply in the nematode or move outside its alimentary tract.

Nematodes act as vectors in the immediate local spread of viruses from infected *Rubus* and weed hosts to healthy *Rubus* plants. Aside from human activities in moving infected planting stock, the long-term, long-distance spread of these viruses is by means of infected seeds, including many weeds. Infected seedlings are symptomless. Transmission of these viruses by pollen to infect the pollinated plant is not known to occur.

Control

Control of AMV, SLRV, RRV, and TBRV can be achieved with certified virus-free planting stock, immune cultivars, vector nematode control, and weed control and crop rotation.

Certified virus-free planting stock. Although overt symptoms of all four of these viruses may occur in *Rubus* cultivars in the field, making roguing possible, frequently there are times of the year and cultivar–virus strain combinations in which diagnostic symptoms are not present in infected *Rubus* plants. Government-sponsored programs for the development and distribution of *Rubus* planting stock free from these and other viruses and their vectors exist in the United Kingdom and other European countries. The use of certified *Rubus* planting stock from virus-free sources is one of the best ways to avoid introducing into the planting area either these nepoviruses or other viruses not already present.

Immune cultivars. In the United Kingdom, some popular red raspberry cultivars (e.g., Glen Isla and Malling Exploit) are immune to the common strains of all four viruses. There are also red raspberry cultivars in the United Kindom that are immune to only one or a few of these nepoviruses. The development and use of immune red raspberry cultivars has reduced the economic damage in the United Kingdom caused by nepoviruses in the commercial red raspberry industry, but resistance-breaking strains of AMV and RRV are also known to occur.

Vector nematode control. Preplant soil fumigation with government-approved nematicides and procedures is commercially used to control vector nematodes in fields to be planted with red raspberry in the United Kingdom and Europe. Effective preplant fumigation is of particular importance in the certified raspberry nursery industry.

Weed control and crop rotation. Because all four European nepoviruses known to infect *Rubus* also infect a wide range of weeds (where they are often seedborne and symptomless), management practices that reduce weeds and their seeds in *Rubus* fields aid in controlling these viruses in *Rubus*. The proper choice of crops for rotation purposes is also important. Populations of *L. elongatus* increase in plantings of clover, grass, and strawberry, so these crops should not immediately precede raspberry in a field rotation.

Selected References

Cadman, C. H. 1956. Studies on the etiology and mode of spread of Scottish raspberry leaf curl disease. J. Hortic. Sci. 31:111-118.

Harrison, B. D., and Murant, A. F. 1977. Nepovirus group. Descriptions of Plant Viruses, No. 185. Commonwealth Mycological Institute and Association of Applied Biologists, Kew, Surrey, England. 4 pp.

Jones, A. T., Mitchell, M. J., and Brown, D. J. F. 1989. Infectibility of some new raspberry cultivars with arabis mosaic and raspberry ringspot viruses and further evidence for variation in British isolates of these two nepoviruses. Ann. Appl. Biol. 115:57-69.

Murant, A. F. 1981a. Nepoviruses. Pages 197-238 in: Handbook of Plant Virus Infections and Comparative Diagnosis. E. Kurstak, ed. Elsevier/North-Holland Biomedical Press, Amsterdam.

Murant, A. F. 1981b. The role of wild plants in the ecology of nematode-borne viruses. Pages 237-248 in: Pests, Pathogens and Vegetation. J. M. Thresh, ed. Pitman, London.

Taylor, C. E., Chambers, J., and Pattullo, W. I. 1965. The effect of tomato black ring virus on the growth and yield of Malling Exploit raspberry. Hortic. Res. 5:19-24.

(Prepared by R. H. Converse)

Tomato Ringspot

Tomato ringspot virus (TomRSV) is the most widespread and important of the nematode-transmitted viruses affecting cultivated *Rubus* in North America. It occurs in *Rubus* and some other crops in South America but is not known to naturally occur outside the Western Hemisphere. Among *Rubus* cultivars, TomRSV causes the most serious disease in red raspberry but also occurs in some cultivated and wild blackberries and raspberry-blackberry hybrids, but not in black raspberry. In red raspberry, damage can range from none (in cultivars where TomRSV is latent) to the production of crumbly fruit to the death of plants severely weakened by the virus.

TomRSV has large natural and experimental host ranges in herbaceous and woody seed plants in more than 35 plant families. Hosts include such fruit crops as strawberry, apple, and peach, in the Rosaceae; blueberry, in the Ericaceae; and many common weeds, such as dandelion (*Taraxacum officinale* Wigg.) and chickweed (*Stellaria media* (L.) Cirillo).

Fig. 28. Chlorotic ring spots in leaf of Malling Exploit red raspberry infected with tomato black ring virus. (Reprinted, by permission, from Taylor et al, 1965; copyright Scottish Crop Research Institute)

Symptoms

In red raspberry, symptom production depends on the cultivar, time of year, and recency of infection. The most striking symptoms occur when the plant is first infected. When shock symptoms appear at the beginning of the growing season, production of foliage and new suckers may be delayed by 1–2 weeks. In some cultivars this new foliage has a distinctive bronze cast, which disappears as the foliage matures (Plate 63). Spring foliage may show yellow rings, line patterns, or vein chlorosis (Plate 64). As the summer progresses, new leaves are usually symptomless, and chronically infected plants often lack distinctive leaf symptoms, although the plants may be dwarfed and weak. Because the virus is vectored by viruliferous dagger nematodes, chronically infected plants usually occur in slowly widening, circular to oval patches in the field. These patches may be mistaken for plants infected by Phytophthora root rot. In the spring, newly infected plants at the edges of such patches will usually develop characteristic symptoms.

The reduction of fruit yield and quality in red raspberry (Plate 65) caused by TomRSV infection varies greatly with the cultivar. The red raspberry cultivar Meeker showed a significant yield reduction in the field when infected by TomRSV but not in drupelet set (crumbly fruit). The reverse was true for the cultivar Puyallup, and neither fruit yield nor drupelet set were significantly depressed in the cultivar Canby.

Most annual weeds are symptomless when naturally infected by TomRSV in the field by seed transmission or by dagger nematode. Several herbaceous greenhouse test plants develop characteristic symptoms when sap-inoculated. Some of the most useful of these include *Chenopodium quinoa* Willd. (Plate 66), cucumber (*Cucumis sativus* L.) (Plate 67), and tobacco (*Nicotiana tabacum* L.) (Plate 68).

In leaves of Himalaya blackberry (*R. procerus* P. J. Müll.), TomRSV caused small chlorotic spots, large yellow blotches at leaflet bases, bright yellow vein chlorosis, and oakleaf patterns. Several blackberry and blackberry-raspberry hybrids graft-inoculated with TomRSV were symptomless.

Causal Agent

TomRSV is a member of the nepovirus group. It has three types of isometric particles 28 nm in diameter, with sedimentation coefficients of 53, 119, and 127 S. The three types of particles contain 0, 40, and 43% single-stranded RNA (ssRNA). Both ssRNA species of this bipartite genome, with relative molecular masses (M_r) of 2.4 and 2.8 $\times$ 10^6 Da, are required for infection. Complementary RNA probes have been made to DNA fragments derived from TomRSV ssRNA. Complementary DNA probes have also been prepared. These probes are able to detect TomRSV in extracts from infected herbaceous hosts, but they give variable results from infected *Rubus* tissues.

Each virion has 60 coat protein subunits of M_r 55,000 Da. TomRSV is often seedborne and is transmitted by several species of the dagger nematode (*Xiphinema* spp.). In the spring, the virus is readily sap-inoculated from new foliage of infected *Rubus* to herbaceous test plants such as cucumber, tobacco, or *Chenopodium quinoa*.

Enzyme-linked immunosorbent assay (ELISA) is commonly used for the detection and identification of TomRSV in *Rubus*. Antisera against several strains of the virus are able to detect it in infected *Rubus* by ELISA. The strongest serological reactions seem to occur in young infected foliage in the spring, but TomRSV has even been detected by ELISA in mature foliage in late summer. Agar gel double diffusion and microprecipitin serological tests have also been successfully used to detect TomRSV in *Rubus* but are less sensitive than ELISA. Commercial sources of polyclonal and monoclonal antiserum are available from organizations such as the American Type Culture Collection. In addition, commercial kits for ELISA detection of TomRSV have been produced (e.g., by Agdia Inc., Elkhart, IN).

Epidemiology

TomRSV is naturally spread by several species of the dagger nematode (*Xiphinema* spp.), including *X. americanum* Cobb and *X. rivesi* Dalmasso. They can transmit this virus for several months. Vector specificity involves specific attachment of TomRSV virions to portions of the lining of the alimentary canal of the nematode. Speciation in this genus is a matter of current study, and other *Xiphinema* spp., but not all, may be vectors. Where vector dagger nematodes are present in a red raspberry field, TomRSV spreads along rows at a rate of approximately 2 m per year. New infections mostly occur next to previously infected raspberry plants, resulting in expanding oval patches of infected plants in a field.

TomRSV is seedborne in *Rubus* and in common weeds, such as chickweed and dandelion. Plants developing from infected seeds are usually symptomless and can serve as foci for new infestations of the virus in fields where dagger nematodes are present.

Control

The major methods of controlling TomRSV in *Rubus* are the following:

1. Using certified planting stock known to be free from TomRSV.
2. Preplant soil testing for the presence of dagger nematodes and preplant fumigating with an approved nematicide if any dagger nematodes are found. Postplanting side dressings with nematicides have not been found to be effective in reducing the spread of TomRSV in fields containing viruliferous nematodes.
3. Roguing infected raspberry plants and additional symptomless plants in each direction beyond the symptomatic plants to eliminate new, latent infections. Studies in Oregon red raspberry fields showed that five symptomless plants in each direction needed to be removed.
4. Controlling weeds to eliminate viruliferous weed reservoirs of TomRSV in the planting.
5. Selecting resistant red raspberry cultivars. No graft-immune ones have been found, but major differences seem to exist among red raspberry cultivars in their resistance to natural TomRSV infection via nematodes. Current research in some laboratories in North America is directed toward incorporating the TomRSV coat protein gene into the red raspberry genome to provide cross-protection against infection.

Selected References

Bitterlin, M. W., and Gonsalves, D. 1988. Serological grouping of tomato ringspot virus isolates: Implications for diagnosis and cross-protection. Phytopathology 78:278-285.

Forer, L. B, and Stouffer, R. F. 1982. *Xiphinema* spp. associated with tomato ringspot virus infection in Pennsylvania fruit crops. Plant Dis. 66:735-736.

Hampton, R., Ball, E., and De Boer, S., eds. 1990. Serological Methods for Detection and Identification of Viral and Bacterial Plant Pathogens. American Phytopathological Society, St. Paul, MN. 389 pp.

Martin, R. R., and Converse, R. H. 1982. An improved buffer for mechanical transmission of viruses from *Fragaria* and *Rubus*. Acta Hortic. 129:69-74.

Stace-Smith, R. 1984. Tomato ringspot virus. Descriptions of Plant Viruses, No. 290. Commonwealth Mycological Institute and Association of Applied Biologists, Kew, Surrey, England. 46 pp.

Stace-Smith, R., and Converse, R. H. 1987. Tomato ringspot virus in *Rubus*. Pages 223-227 in: Virus Diseases of Small Fruits. R. H. Converse, ed. U.S. Dep. Agric. Agric. Handb. 631.

Taylor, C. E., and Robertson, W. M. 1975. Acquisition, retention and transmission of viruses by nematodes. Pages 253-276 in: Nematode Vectors of Plant Viruses. F. Lamberti, C. E. Taylor, and J. W. Seinhorst, eds. NATO Advanced Study Institute, Series A: Life Sciences. Vol. 2. Plenum Press, New York.

(Prepared by R. H. Converse)

Pollen-Transmitted Diseases

Raspberry Bushy Dwarf

Raspberry bushy dwarf virus (RBDV) occurs naturally worldwide in many *Rubus* species and cultivars. In North America, it naturally infects many red raspberry, black raspberry, and blackberry-raspberry hybrid cultivars. RBDV also occurs in wild *R. idaeus* L. var. *strigosus* (Michx.) Maxim., *R. occidentalis* L., *R. parviflorus* Nutt., and *R. leucodermis* Douglas ex Torr & A. Gray. Experimental hosts (by graft inoculation) include *Cydonia oblonga* Mill. 'C7/1,' *Fragaria vesca* L., and seven *Rubus* spp. and (by sap inoculation) *Chenopodium quinoa* Willd. and several other *Chenopodium* spp., *Cucumis sativus* L., *Nicotiana clevelandii* A. Gray, *Phaseolus vulgaris* L., and plants in some other genera. *C. quinoa* is often used for the detection of RBDV by sap inoculation. *R. molaccanus* L. and *C. oblonga* 'C7/1' develop diagnostic symptoms when graft-inoculated with RBDV.

In a British Columbia field study, the red raspberry cultivars Canby and Meeker showed significant reductions in cane height (22%), cane diameter (14%), and fruit yield (72%) compared to uninoculated controls. In an Oregon field study of the black raspberry cultivar Munger, RBDV-infected plants had a significant loss of vigor (a 38% decrease in the number of primocanes and a 34% decrease in the weight of floricanes) compared to uninoculated controls, but there were no significant differences in fruit yield.

RBDV has not been reported to occur in the field in eastern North American blackberry cultivars, but it sometimes occurs in Pacific Coast blackberry cultivars. In the United States, it is common in some blackberry-raspberry hybrids, such as the cultivar Boysen in California. This virus is the major cause of disease in New Zealand and is common in Europe.

Symptoms

Although pollen-borne, RBDV does not cause pollen abortion but does cause drupelet abortion, which leads to crumbly fruit in some red raspberry cultivars. RBDV is symptomless in many North American red raspberry cultivars. In some Pacific Coast red raspberry cultivars, the virus may cause foliar ring and line patterns resembling those caused by several nepoviruses (Plate 69). In other cases it causes interveinal chlorosis, which produces a condition known as yellows. Mixed infections of RBDV with other viruses may cause more severe damage to red raspberry cultivars than single infections. In combination with black raspberry necrosis virus, RBDV causes dwarfing and shoot proliferation in red raspberry, a typical bushy dwarf condition.

Most Pacific Coast trailing blackberry (*R. ursinus* Cham. & Schlechtend.) cultivars and their hybrids that were studied in Scotland were symptomless when infected with RBDV, except the cultivar Marion, which developed yellows. Blackberry-raspberry hybrids, such as boysenberry, appear to be symptomless when naturally infected with RBDV in the United States. A discrepancy occurs in the symptoms and damage caused by RBDV in the blackberry-raspberry hybrid loganberry in the United Kingdom and in the United States. In the United Kingdom, highly significant reductions occurred in fruit yield (36%) and cane weight (34%) in loganberries infected with RBDV. The disease is called loganberry degeneration. It has been shown in Scotland to be caused by red raspberry isolates of RBDV, and the virus causing it is serologically related to RBDV. In western United States, RBDV has been detected in symptomless, apparently vigorous loganberry.

RBDV can be mechanically transmitted to several herbaceous hosts, but several test plants in the genus *Chenopodium*, particularly *C. quinoa* Willd., are commonly used as test hosts. In *Rubus*, the virus is most readily sap-transmitted from young, vigorously growing leaves in the spring to young, rapidly growing *C. quinoa* plants. Initial symptoms of RBDV on *C. quinoa* within 10 days of mechanical inoculation are occasional small, chlorotic local lesions, followed by systemic small chlorotic leaf spots and mosaic and ring and line patterns (Plate 70). No necrosis is associated with RBDV infection of *C. quinoa*—a major difference from most other sap-transmissible viruses from *Rubus*.

Causal Agent

The RBDV virion is quasi-isometric (33 nm), with a sedimentation coefficient ($s_{20,w}$) of 115 S. RBDV has not been assigned to a recognized plant virus group. The coat protein subunit has a relative molecular mass (M_r) of 29,000 Da. The virus has three species of single-stranded RNA of M_r 2.2, 0.9, and 0.4×10^6 Da.

Polyclonal and monoclonal antibodies have been produced against RBDV. Antisera against red raspberry isolates of RBDV recognize this virus in all hosts tested, but RBDV from black raspberry is a distinct serotype. Enzyme-linked immunosorbent assay (ELISA) is now the method of choice for the detection of the virus.

Although not serologically distinguishable from standard isolates of RBDV, resistance-breaking (RB) isolates of the virus have been found in the United Kingdom that are capable of infecting the raspberry cultivars Glen Clova, Malling Admiral, Malling Delight, and Malling Jewel, which are immune to the common strain of RBDV. The red raspberry cultivar Willamette, widely grown in the Pacific Northwest, is immune to the common strain of RBDV. The RB strain has not yet been reported to occur in North America. Willamette is susceptible to the RB strain in tests in the United Kingdom. Of the European and North American red raspberry and blackberry cultivars tested in the United Kingdom, only three were immune to RBDV-RB by graft inoculation (see Control).

Epidemiology

RBDV is seedborne in *R. idaeus*, *R. phoenicolasius* Maxim, and *R. parviflorus*. In red raspberry, up to 77% seed transmission occurred when the pollen parent and the seed parent were infected. Homogenized, dormant, infected black raspberry seed is a convenient reference source of RBDV for ELISA.

The virus is pollen-associated in all infected *Rubus* spp. studied and in *F. vesca*. In serological studies with pollen grains of *R. parviflorus*, RBDV was found to occur both on, and probably in, pollen grains. RBDV spread by pollen can infect the pollinated *Rubus* plant as well as resulting seed. RBDV infection of growing plants is believed to occur by the movement of viruliferous pollen to healthy plants. Deflowered red and black raspberry plants do not become infected with the virus, but plants allowed to flower normally become infected within two or three flowering seasons if located near RBDV-infected plants.

No vectors have been shown to be directly involved in the infection of *Rubus* plants by RBDV. However, *Rubus* flowers are commonly visited by pollen- and nectar-feeding insects, including honey bees, and these insects doubtless play a major role in the efficient movement of viruliferous pollen from RBDV-infected to healthy *Rubus* plants.

Control

The use of immune cultivars is the only effective means of controlling RBDV. The red raspberry cultivars Willamette, Haida, Heritage, and Latham are graft-immune to the common isolates of RBDV tested in North America and the United Kingdom, but all black raspberry, blackberry, and blackberry-raspberry hybrid cultivars tested were graft-susceptible. Because most *Rubus* cultivars are symptomless when RBDV-infected, it is important that plant breeders be aware of the presence of the virus in potential parental *Rubus* lines.

All *Rubus* cultivars graft-inoculated in the United Kingdom with RBDV-RB were susceptible except for the cultivars Haida,

Preussen, and Rannaya Sladkaya. In studies with Haida, this graft immunity was found to be heritable. The gene *Bu* and a second multigenic component appear to be responsible for this graft immunity to RBDV-RB. The red raspberry cultivars Heritage and Malling Promise are difficult to graft-inoculate with RBDV-RB and may represent additional genetic resources for RBDV-RB resistance.

In cases where it is necessary to eliminate RBDV from a *Rubus* clone, thermotherapy and shoot tip culture can be used. RBDV is rather resistant to in vivo heat inactivation. An isolate in the black raspberry cultivar Munger was eliminated in 50% of plants propagated from heated shoot tips from source plants held for 64 days at a constant 37°C. In red raspberry (cultivar not specified) RBDV was eliminated from 50% of plants propagated from heated shoot tips from source plants held for 72 days at temperatures from 33 to 42°C.

Regardless of virus content, shoot tips from heat-treated red raspberry plants rooted much more readily than comparable shoot tips from plants grown at ambient temperatures.

Selected References

Bulger, M., Stace-Smith, R., and Martin, R. R. 1990. Transmission and field spread of raspberry bushy dwarf virus. Plant Dis. 74:514-517.

Credi, R., Shier, J. L., and Stace-Smith, R. 1986. Occurrence of raspberry bushy dwarf virus in native thimbleberry in British Columbia. Acta Hortic. 186:17-22.

Daubeny, H. A., Freeman, J. A., and Stace-Smith, R. 1982. Effect of raspberry bushy dwarf virius on yield and cane growth in susceptible red raspberry cultivars. HortScience 17:645-657.

Jennings, D. L., and Jones, A. T. 1989. Further studies on the occurrence and inheritance of resistance in red raspberry to a resistance-breaking strain of raspberry bushy dwarf virus. Ann. Appl. Biol. 114:317-323.

Jones, A. T., Murant, A. F., Jennings, D. L., and Wood, G. A. 1982. Association of raspberry bushy dwarf virus with raspberry yellows disease; reaction of *Rubus* species and cultivars, and the inheritance of resistance. Ann. Appl. Biol. 100:135-147.

Kurppa, A., and Martin, R. R. 1986. Use of double-stranded RNA for detection and identification of virus diseases of *Rubus* species. Acta Hortic. 186:51-62.

Martin, R. R. 1984. Monoclonal antibodies define three different antigenic regions on raspberry bushy dwarf virus. (Abstr.) Can. J. Plant Pathol. 6:265.

Murant, A. F. 1976. Raspberry bushy dwarf virus. Descriptions of Plant Viruses, No. 165. Commonwealth Mycological Institute and Association of Applied Biologists, Kew, Surrey, England. 4 pp.

Murant, A. F. 1987. Raspberry bushy dwarf. Pages 229-234 in: Virus Diseases of Small Fruits. R. H. Converse, ed. U.S. Dep. Agric. Agric. Handb. 631.

(Prepared by R. H. Converse)

Viral Diseases with Unknown Methods of Natural Spread

Apple Mosaic

Apple mosaic virus (ApMV), a member of the ilarvirus group, naturally infects *Betula, Aesculus, Humulus,* and several crop genera in the family Rosaceae (*Malus, Prunus,* and *Rosa*). ApMV was first reported in *Rubus* in several blackberry and raspberry cultivars in the United States and subsequently in red raspberry cultivars in Germany and in *R. parviflorus* Nutt. (thimbleberry) in British Columbia. To date, there have been no studies of the effects of ApMV on fruit yield in the genus. Infected *Rubus* plants are often symptomless and generally appear to produce fruit of normal yield and quality. In the red raspberry cultivar Schoenemann naturally infected in Germany, ApMV caused bright yellow-white leaf mosaic and sometimes pale green foliar rings and spots (Plate 71) but no fruit symptoms.

ApMV has a large experimental host range when transmitted by grafting or sap inoculation. Some herbaceous greenhouse test plants that are readily sap-inoculated by ApMV and produce good but not diagnostic leaf symptoms include *Chenopodium quinoa* Willd., cucumber (*Cucumis sativus* L. 'Straight Eight'), bean (*Phaseolus vulgaris* L. 'Pinto' and 'Black Turtle'), and cowpea (*Vigna unguiculata* (L.) Walp.).

The thimbleberry isolate of ApMV was recently purified in Canada from sap-inoculated *C. quinoa*. This isolate sedimented in two bands with sedimentation coefficients of 82 and 100 S and contained four single-stranded RNA species with relative molecular masses (M_r) of 1.18, 0.99, 0.72, and 0.34×10^6 Da. The coat protein subunit had an M_r of 30,000 Da. These data are similar to those reported for isolates of ApMV from apple.

Antisera prepared against the virus from rose and apple have been successfully used in Oregon to detect ApMV in *Rubus* by enzyme-linked immunosorbent assay (ELISA). In Canada, both rabbit polyclonal antiserum and murine monoclonal antibodies have been produced against the thimbleberry isolate. These antisera detected ApMV isolates from thimbleberry and apple that showed a reaction of complete homology in agar gel double diffusion tests. In double-antibody sandwich ELISA tests, ApMV-infected thimbleberry sap gave readings 10 times greater than those of comparable uninfected thimbleberry sap. ELISA is currently the method of choice for detection of ApMV in *Rubus*.

Natural spread of the virus in rosaceous crops by vectors is unknown. In some plants in which natural root grafting is common (such as apple), ApMV is believed to spread by this means, but no evidence has yet been presented.

Certified *Rubus* planting stock from source plants free of ApMV should provide adequate control of this virus. Recent studies in Germany have shown that the virus can be eliminated relatively easily from infected red raspberry cultivars by excision, tissue culture, and regeneration of shoot apices 1–3 mm long from axillary buds.

Selected References

Baumann, G., Casper, R., and Converse, R. H. 1982. The occurrence of apple mosaic virus in red and black raspberry and blackberry cultivars. Acta Hortic. 129:13-20.

Baumann, G., Casper, R., and Converse, R. H. 1987. Apple mosaic virus in *Rubus*. Pages 246-248 in: Virus Diseases of Small Fruits. R. H. Converse, ed. U.S. Dep. Agric. Agric. Handb. 631.

Fulton, R. W. 1972. Apple mosaic virus. Descriptions of Plant Viruses, No. 83. Commonwealth Mycological Institute and Association of Applied Biologists, Kew, Surrey, England. 4 pp.

Stace-Smith, R., and Shier, J. L. 1988. Some properties of apple mosaic virus isolated from thimbleberry in British Columbia. Acta Hortic. 236:73-80.

Theiler-Hedtrich, R., and Baumann, G. 1989. Elimination of apple mosaic virus and raspberry bushy dwarf virus from infected red raspberry (*Rubus idaeus* L.) by tissue culture. J. Phytopathol. 127:193-199.

(Prepared by R. H. Converse)

Blackberry Calico

Blackberry calico virus (BCV) occurs in wild Pacific Coast trailing blackberry (*Rubus ursinus* Cham. & Schlechtend.) and several Pacific Coast trailing blackberry cultivars, such as Chehalem, Marion, Olallie, and Waldo, and blackberry-raspberry hybrids, such as the cultivars Boysen, Logan, and the mutant Thornless Logan. Field-run stocks of Chehalem and Thornless Logan appear to be almost universally infected. In the Netherlands, BCV has occurred naturally in the blackberry (*R. laciniatus* Willd.) cultivar Thornless Evergreen, and *Nicotiana occidentalis* Wheeler subsp. *obliqua* accession P-1c was successfully mechanically inoculated with it, developing small chlorotic-necrotic lesions on the inoculated leaves.

Cultivars such as Marion and Thornless Logan infected with BCV appear to flower normally and bear horticulturally acceptable fruit crops. However, a recent field comparison in Oregon of healthy Marion blackberry and Marion infected with BCV showed that the virus significantly depressed fruit yield (by 14%) and the number and vigor of primocanes (by 39 and 32%, respectively). In hot, dry weather some leaves with severe calico symptoms may drop in unirrigated BCV-infected blackberry plants.

A graft-transmissible, brilliant yellow veinclearing in red raspberry, earlier referred to as raspberry calico, is probably a symptom of raspberry bushy dwarf virus. Various genetically induced chloroplast abnormalities in *Rubus* cause blotchy, chlorotic leaf sectors that may resemble blackberry calico but are not graft-transmissible.

Symptoms

There are no known fruit or cane symptoms. Field symptoms in BCV-infected blackberry or blackberry-raspberry hybrid cultivars appear first on floricane leaves at flowering time. Symptoms vary from blotchy chlorosis and discreet, small yellow spots to yellow ring and line patterns and oakleaf patterns (Plate 72). On primocanes, a similar range of symptoms develops on mature leaves during hot summer weather. Young leaves not fully expanded are symptomless, and symptom production is very uneven among primocanes and floricanes of an infected plant and along individual canes. Once they have developed calico symptoms, leaves continue to show symptoms for the rest of the growing season. In some instances leaf chlorosis becomes severe, leading to red pigmentation, bleaching, and necrosis. Symptoms may fail to appear on infected plants during cool, cloudy weather.

Causal Agent

In the Netherlands BCV was sap-transmitted from Thornless Evergreen with leaf pattern symptoms resembling those of blackberry calico disease in the western United States to *N. occidentalis* accession P-1c. A flexuous, rod-shaped virus with a modal length of 627 nm was purified from infected *N. occidentalis*. An antiserum prepared in the Netherlands against this virus reacted in Oregon with leaf samples (with typical calico disease symptoms) of blackberry cultivars Chehalem, Marion, and Waldo and also the hybrid cultivar Thornless Logan in double-antibody sandwich enzyme-linked immunosorbent assay (ELISA) tests. On the basis of its morphology and its positive serological reactions with antibodies against several carlaviruses, BCV is considered a member of the PVS subgroup of the carlavirus group.

BCV can be experimentally transmitted to Marion blackberry by approach graft, but symptom development in the greenhouse is very slow (6–24 months). Graft analysis revealed a minor difference in experimental host range between BCV isolates from the cultivar Chehalem and those from the cultivars Marion and Thornless Logan. Tobacco streak virus (TSV), which frequently occurs in symptomless blackberry plants along the Pacific coast, causes a necrotic shock reaction when grafted into the cultivar Marion; this makes it unsuitable for graft analysis for latent BCV in plants also infected with TSV. Loganberry seedlings can serve as indicators for BCV in the presence of TSV by graft analysis. Serological detection of BCV using ELISA is probably the fastest, most sensitive technique available for detection of latent BCV.

In recent experiments in the United Kingdom, the blackberry cultivars Marion and Olallie were graft-inoculated with wineberry latent virus and raspberry bushy dwarf virus together, causing symptoms typical of calico disease. Graft inoculation of Marion and Olallie with wineberry latent virus alone was not done. In the United States, calico disease often occurs in these hosts in the absence of raspberry bushy dwarf virus. The relationship of wineberry latent virus to BCV remains to be determined.

Epidemiology

No vector is known to spread BCV. The virus appears to spread very slowly in the field. In Oregon, some fields of the cultivar Marion 20–29 years old have been observed, with 5–36% of the plants showing typical calico symptoms. In Marion, BCV appears to be unevenly distributed among canes and within canes of an infected plant. One third of nodal propagants from calico-infected Marion canes in a field test in Oregon grew out to be symptomless plants, some of which indexed negative for BCV by ELISA.

Control

Generally, growers and nurserymen seem to have disregarded this disease when propagating infected cultivars. The lack of severe damage caused by BCV and its apparent uneven distribution within infected plants may account for its widespread occurrence along the Pacific Coast of Oregon and California without completely infecting old fields.

BCV was not eliminated from Thornless Loganberry plants heat-treated at 37°C for 17 days, but it was eliminated in Canada from clones of the cultivar by shoot tip propagation after exposure to a constant 37°C for 35 days.

Selected References

Converse, R H. 1987. Blackberry calico. Pages 245-246 in: Virus Diseases of Small Fruits. R. H. Converse, ed. U.S. Dep. Agric. Agric. Handb. 631.

Converse, R. H., and Kowalczyk, K. L. 1980. Graft transmission of calico from three blackberry cultivars. Acta Hortic. 112:47-52.

Jones, A. T., Mitchell, M. J., McGavin, W. J., and Roberts, I. M. 1991. Further properties of wineberry latent virus and evidence for its possible involvement in calico disease. Ann. Appl. Biol. 117:571-581.

van der Meer, F. A. 1989. *Nicotiana occidentalis*, a suitable test plant in research on viruses of small fruit crops. Acta Hortic. 236:27-35.

Wieczorek, A., and Stace-Smith, R. 1989. A conserved region in carlaviruses detected by a monoclonal antibody. (Abstr.) Phytopathology 79:912.

(Prepared by R. H. Converse)

Cherry Leaf Roll

Cherry leaf roll virus (CLRV) was first reported in England in *Rubus procerus* P. J. Müll. 'Himalaya Giant,' in which it was sometimes lethal. It was subsequently reported in cultivated red raspberry in New Zealand, where infected red raspberry plants were depressed in vigor and exhibited severe leaf symptoms.

The virus has not been reported on wild or cultivated *Rubus* in North America, but it has a wide natural host range in wild and cultivated trees and shrubs in Europe and North America.

Symptoms

In the blackberry cultivar Himalaya Giant naturally infected in England, CLRV is reported to cause chlorotic mottling and line patterning in leaves. Infected plants are stunted and may be killed.

In three red raspberry cultivars naturally infected by CLRV in New Zealand, infected plants had stunted, distorted leaves with severe chlorotic mottle and ring and line patterns (Plate 73). Resistance or immunity to CLRV in *Rubus* has not been reported.

A large number of herbaceous hosts can be experimentally infected by *Rubus* isolates of CLRV by mechanical inoculation. Among such susceptible hosts are *Chenopodium amaranticolor* Coste & Reyn., *C. quinoa* Willd., *C. foetidum* Schrad., *Cucumis sativus* L., and *Nicotiana tabacum* L. 'Xanthi-nc'. None of the symptoms produced on these hosts is diagnostic for CLRV.

Causal Agent

CLRV is a single-stranded RNA, isometric virus, 28 nm in diameter, with a bipartite genome. It is similar to nepoviruses in its three classes of virus particles with sedimentation coefficients of 128, 115, and 54 S. The two single-stranded RNAs have relative molecular masses (M_r) of 2.4 and 2.1×10^6 Da, encapsidated in the 128 and 115 S particles; the 54 S particle is RNA-free. There is only one virus coat protein subunit, of M_r 54,000 Da.

A number of antisera against CLRV isolates from different hosts have been prepared, and although cross-recognizable, they tend to be somewhat serologically distinct according to host source. The raspberry isolates of CLRV in New Zealand were all detected by antisera against CLRV from American dogwood (*Cornus nuttallii* Audubon). Detection of CLRV in *Rubus* is probably best done by enzyme-linked immunosorbent assay serology, with the recognition that because of wide differences in serological specificity among serotypes, antisera against a number of CLRV serotypes should be used before concluding that a suspect *Rubus* plant is free of CLRV.

Epidemiology

No vector is commonly associated with natural spread of CLRV. Early claims of transmission by *Xiphinema* spp. were later explained as probable direct virus contamination in the test pots. In walnut, CLRV has been found to be spread by pollen. Whereas this mode of transmission has not been reported in *Rubus*, it remains a distinct possibility, as does seed transmission, which is also common in many infected plants but has not yet been tested in *Rubus*.

Control

In New Zealand, certified red raspberry planting stock known to be free from CLRV is available. The virus has not been recently detected in New Zealand, and currently no control measures are recommended for it there. If CLRV is shown to be spread by viruliferous red raspberry pollen, the eradication of infected stock would become a necessary part of control procedures where this virus appeared in commercial *Rubus* plantings. No attempts have been made to rid *Rubus* clones of CLRV by heat therapy or shoot apex culture. The virus was eliminated by shoot apex culture from rhubarb, however, and presumably similar methods could be used as needed to free *Rubus* clones of this virus.

Selected References

Cooper, J. I. 1979. Virus Diseases of Trees and Shrubs. Institute of Terrestrial Ecology, Cambridge. 74 pp.

Jones, A. T. 1985. Cherry leafroll virus. Descriptions of Plant Viruses, No. 306. Commonwealth Mycological Institute and Association of Applied Biologists, Kew, Surrey, England. 6 pp.

Jones, A. T. 1987. Cherry leaf roll virus in *Rubus*. Pages 220-223 in: Virus Diseases of Small Fruits. R. H. Converse, ed. U.S. Dep. Agric. Agric. Handb. 631.

Jones, A. T., McElroy, F. D., and Brown, D. J. F. 1981. Tests for transmission of cherry leaf roll virus using *Longidorus*, *Paralongidorus* and *Xiphinema* nematodes. Ann. Appl. Biol. 99:143-150.

Jones, A. T., and Wood, G. A. 1979. The virus status of raspberries (*Rubus idaeus* L.) in New Zealand. N.Z. J. Agric. Res. 22:173-182.

Mircetich, S. M., Sanborn, R. R., and Ramos, D. E. 1980. Natural spread, graft transmission and possible etiology of walnut blackline disease. Phytopathology 70:962-968.

Stace-Smith, R., and Ramsdell, D. C. 1987. Nepoviruses of the Americas. Pages 131-166 in: Current Topics in Vector Research. Vol. 3. K. F. Harris, ed. Springer-Verlag, New York.

(Prepared by R. H. Converse)

Tobacco Streak

The *Rubus* strain of tobacco streak virus (TSV-R) was first reported in California in 1966 in several blackberry and blackberry-raspberry cultivars. It was later found to be common in the Pacific Northwest in cultivated black raspberry and wild and cultivated *R. ursinus* Cham. & Schlechtend., the Pacific Coast trailing blackberry, and its hybrids. TSV-R is not known to naturally occur outside the genus *Rubus* but has been found in this genus in North America and Australia. No *Rubus* spp. have been found that are graft-immune to TSV-R. There are several strains of tobacco streak virus. Some, such as the white clover, bean red node, and HF type strains, have been isolated from many non-*Rubus* crops and wild plant species in many parts of the world.

Rubus cultivars infected with TSV-R are symptomless and produce acceptable yields. In one Oregon field test, primocane production of the Pacific Northwest blackberry cultivar Santiam was significantly (32%) greater in virus-tested plants than in plants infected by TSV-R. In Scotland, spring foliation was delayed in one red raspberry cultivar when it was infected with TSV-R. This virus often occurs in complex with other viruses, such as raspberry bushy dwarf virus, but data are lacking on the effect of TSV-R on *Rubus* growth or yield in complex with other viruses.

Symptoms

Natural infections of TSV-R in *Rubus* are usually symptomless. Some *Rubus* hosts, such as *R. henryi* Hemsl. & Kuntze, *R. phoenicolasius* Maxim., *R. procerus* P. J. Müll., the blackberry-raspberry hybrid cultivar Logan, and the Pacific Northwest trailing blackberry cultivars Olallie and Marion, exhibit hypersensitive necrotic reactions at the graft union when graft-inoculated with TSV-R. Other *Rubus* spp., such as red and black raspberry and the blackberry-raspberry hybrid cultivar Boysen, remain symptomless when graft-inoculated with this virus.

TSV-R resembles other strains of TSV in its mechanical transmissibility to a number of common greenhouse herbaceous indicator plants. In *Nicotiana tabacum* L. 'Haranova' TSV-R produces necrotic local lesions on inoculated leaves within 1 week after inoculation when young, actively growing leaves of a *Rubus* plant infected with TSV-R (Plate 74) are used. *Chenopodium quinoa* Willd. is also susceptible to TSV-R infection; it develops necrotic local lesions on inoculated leaves within a few days after sap inoculation. Later, the virus usually becomes systemic, causing death of shoot tips. The symptoms caused by TSV in sap-inoculated herbaceous hosts are not diagnostic, because many other viruses produce similar symptoms when transmitted to these hosts by sap inoculation.

Causal Agent

TSV-R is a strain of tobacco streak virus, the type member of the ilarvirus group, which is a group of single-stranded RNA (ssRNA) viruses with tripartite genomes. TSV-R virus particles are quasi-isometric, averaging 28 nm in diameter. During sucrose rate–zonal density gradient centrifugation, these particles separate into three bands with sedimentation coefficient values of 90, 98, and 113 S. The coat protein subunit of this virus has a relative molecular mass (M_r) of 28,700 Da, and TSV-R contains four ssRNA species of M_r 1.1, 0.9, 0.7, and 0.3×10^6 Da. In order to be infective, either all four ssRNAs or all but the smallest plus coat protein are required. Complementary DNA probes to the strawberry necrotic shock strain of TSV (TSV-SNS) that have about the same sensitivity as TSV-SNS antisera in enyzyme-linked immunosorbent assay (ELISA) for TSV-SNS and TSV-R have been developed.

Antisera have been prepared against TSV-R and several other TSV serotypes. The TSV-R serotype is serologically similar to the SNS serotype. Serologically, it is more distantly related to the bean red node strain of TSV and only slightly related serologically to the white clover strain. ELISA is a rapid and reliable method of detecting TSV-R in *Rubus*, but antisera prepared against TSV-R or TSV-SNS should be used in this assay.

Black raspberry latent virus resembles TSV-R in many properties but is serologically distinct from it.

Epidemiology

TSV-R is pollen- and seed-transmitted in *Rubus* and probably enters into breeders' crosses by these means. Some strains of TSV are known to be vectored by thrips (*Frankliniella occidentalis* (Pergande) and *Thrips tabaci* Lindeman). Although TSV-R is strongly but not totally flower-associated in its field spread in black raspberry, there are no published data on the ability of thrips to transmit TSV-R, with or without the involvement of viruliferous *Rubus* pollen. The method of spread of TSV-R in the field in *Rubus* is unknown.

The virus is so unevenly distributed in the shoots of *R. ursinus* 'Santiam' grown at normal greenhouse temperatures that 55% of plants propagated from three-node softwood cuttings from plants infected with TSV-R were found to be free of this virus.

Control

The control of TSV-R is primarily through the use of certified stock from mother plants that index free of this virus. Floricane removal before flowering is a wise precaution to reduce possible flower-associated spread of TSV-R in nursery plantings of blackberry and black raspberry in areas where the virus is common. Red raspberry is not as frequently infected by TSV-R as are the blackberry-raspberry hybrid cultivars Boysen and Logan, black raspberry, or the wild Pacific trailing blackberry and its cultivars. In one 5-year Oregon field test, the annual infection rate of TSV-R in the red raspberry cultivar Willamette was 7% in plants in a row immediately next to a row of infected boysenberry compared to 1 and 0.5% in plants two and three rows away, respectively, from the infected boysenberry row. Unlike black raspberry or boysenberry, red raspberry was equally likely to be infected whether flowers were present or not. To reduce natural spread of TSV-R, healthy red raspberry plantings should probably not be made adjoining plantings of boysenberry, loganberry, or blackberry cultivars infected with TSV-R.

Selected References

Converse, R. H. 1978. Uneven distribution of tobacco streak virus in Santiam blackberry before and after heat therapy. Phytopathology 68:241-244.

Converse, R. H. 1986. Rates and patterns of tobacco streak virus spread in Boysen and in red raspberry. Acta Hortic. 186:31-37.

Converse, R. H., and Bartlett, A. B. 1979. Occurrence of viruses in some wild *Rubus* and *Rosa* species in Oregon. Plant Dis. Rep. 63:441-444.

Converse, R. H., and Kowalczyk, K. L. 1980. Graft transmission of calico from three blackberry cultivars. Acta Hortic. 112:47-52.

Fulton, R. W. 1971. Tobacco streak virus. Descriptions of Plant Viruses, No. 44. Commonwealth Mycological Institute and Association of Applied Biologists, Kew, Surrey, England. 4 pp.

Jones, A. T., and Mayo, M. A. 1975. Further properties of black raspberry latent virus and evidence for its relationship to tobacco streak virus. Ann. Appl. Biol. 79:297-306.

Kaiser, W. J., Wyatt, S. D., and Pesho, S. D. 1982. Natural hosts and vectors of tobacco streak virus in eastern Washington. Phytopathology 72:1508-1512.

Sdoodee, R., and Teakle, D. S. 1987. Transmission of tobacco streak virus by Thrips tabaci: A new method of plant virus transmission. Plant Pathol. 36:377-380.

Stace-Smith, R. 1987. Tobacco streak virus in *Rubus*. Pages 235-239 in: Virus Diseases of Small Fruits. R. H. Converse, ed. U.S. Dep. Agric. Agric. Handb. 631.

Stenger, D. C., Mullin, R. H., and Morris, T. J. 1987. Characterization and detection of the strawberry necrotic shock isolate of tobacco streak virus. Phytopathology 77:1330-1337.

(Prepared by R. H. Converse)

Wineberry Latent Virus

Wineberry latent virus (WLV) was discovered in a single symptomless plant of wineberry, *Rubus phoenicolasius* Maxim., which was growing in an experimental planting in Scotland. The plant originated in the United States, where wineberry is established in the wild in the Northeast. The species is native to Japan. Experimentally, WLV can be graft-transmitted to some wild *Rubus* spp. and to some red raspberry, black raspberry, blackberry, and blackberry-raspberry hybrid cultivars. Other red raspberry and blackberry cultivars appeared to be graft-immune. A number of commonly used herbaceous virus indicator plants, including *Catharanthus roseus* (L.) G. Don, several *Chenopodium* spp., *Gomphrena globosa* L., and *Lycopersicon esculentum* Mill., can be successfully sap-inoculated with the virus by using a buffer containing 2% nicotine alkaloid. Recent studies in Scotland have suggested that WLV might be the cause of blackberry calico disease. If this relationship is confirmed in further studies, the economic importance of WLV will have been established.

Symptoms

The blackberry cultivars Marion and Olallie developed calico symptoms when graft-inoculated in Scotland with a mixture of WLV and raspberry bushy dwarf virus (RBDV). WLV causes small local lesions on sap-inoculated leaves of several *Chenopodium* spp. In *C. amaranticolor* Coste & Reyn., these lesions become large and necrotic after 3 weeks (Plate 75).

Causal Agent

The virion of WLV is a flexuous rod, about 620×12 nm, and is serologically unrelated to several potexviruses, carlaviruses, capilloviruses, or closteroviruses. Partially purified WLV contained a single species of single-stranded RNA of M_r 2.78×10^6 Da and a coat protein subunit of M_r 31,000 Da. In nature, WLV was found only in a single symptomless plant (in which it was associated with RBDV), from which it was experimentally separated by sap inoculation of tomato (which is immune to RBDV). Subsequently, partial purification of WLV from *C. amaranticolor* was accomplished, yielding small quantities of virus that had a strong tendency to be fragmented and aggregated end-to-end. Antiserum against WLV has been prepared.

Epidemiology and Control

WLV was not experimentally transmitted by the aphid *Macrosiphum euphorbiae* (Thomas) and is not seedborne in wineberry. Natural spread is unknown. No control measures have been developed for this virus in wineberry. Unexpectedly, some red raspberry cultivars known to be resistant to graft inoculation by RBDV were also found to be resistant to graft inoculation by WLV.

Selected References

Jones, A. T. 1977. Partial purification and some properties of wineberry latent, a virus obtained from *Rubus phoenicolasius*. Ann. Appl.

Biol. 86:199-208.
Jones, A. T. 1987. Wineberry latent virus. Pages 239-241 in: Virus Diseases of Small Fruits. R. H. Converse, ed. U.S. Dep. Agric. Agric. Handb. 631.
Jones, A. T., Mitchell, M. J., McGavin, W. J., and Roberts, I. M. 1991. Further properties of wineberry latent virus and evidence for its possible involvement in calico disease. Ann. Appl. Biol. 117:571-581.

(Prepared by R. H. Converse)

Other Viruses and Viruslike Agents

Alpine Mosaic Agent

In Oregon studies, a graft-transmissible agent of presumed but unproven viral etiology was found to be present in some symptomless clones of the erect blackberry cultivar Darrow. This agent caused faint leaf mottling symptoms when leaflets from infected Darrow plants were grafted into seedlings of *Fragaria vesca* L. var. *semperflorens* (Duch.) Ser. 'Alpine' (Plate 76). No sap-transmissible viruses were obtained from the infected Darrow or Alpine plants, and the identity and relationships of the causal agent, called Alpine mosaic agent, are unknown. The agent was found to be present in some Darrow clones that set normal fruit as well as some that were totally sterile. The sterility condition was not graft-transmissible to fruitful Darrow and is thought to be of genetic origin, although there is a report from Missouri of graft-transmissible sterility in blackberry.

Bean Yellow Mosaic Virus

The bean yellow mosaic virus, a potyvirus, was isolated once from two red raspberry plants in New York State. No information is available on the symptoms it causes in red raspberry.

Black Raspberry Latent Virus

Black raspberry latent virus (BRLV), an isometric virus, was found to occur symptomlessly in cultivated black raspberry in the eastern United States and to be seed-transmitted and also pollen-transmitted to the pollinated plant. The first report of BRLV noted that it was serologically unrelated to other known viruses, but a subsequent report with plant material imported to Scotland equated BRLV with tobacco streak virus. Several virus isolates subsequently collected from old cultivated black raspberry fields in Maryland and Pennsylvania were sap-transmissible to herbaceous hosts, such as *Chenopodium quinoa* Willd., to produce symptoms similar to those caused by BRLV and the *Rubus* strain of tobacco streak virus (TSV-R). These isolates did not react in enzyme-linked immunosorbent assays with antisera against an isolate of TSV from cultivated black raspberry in Oregon. The possibility of pollen infection of the pollinated plant by BRLV, if confirmed, would distinguish this virus from TSV-R both biologically and in terms of its required control measures.

Black Raspberry Streak

The development of faint, water-soaked, vertical lines on the surface of young black raspberry primocanes is the diagnostic symptom of streak disease of cultivated black raspberry (*Rubus occidentalis* L.). In the case of mild streak disease, the streaks are easily obliterated by rubbing the bloom off young canes. The more pronounced streaks associated with severe streak disease are illustrated in Plate 77. The large bluish streaks (*bluestem*) caused by the Verticillium wilt fungus on red raspberry canes are much wider, more pronounced, and more permanent than those of the streak diseases. Leaflets on black raspberry plants infected with streak disease remain a normal green color but may hook downward along the midrib, and berries from infected plants may be smaller and less glossy than those from healthy plants.

Black raspberry streak disease should not be confused with tobacco streak virus, which is symptomless and common in black raspberry. Black raspberry streak disease occurs in the northeastern United States and is one of the least understood of the graft-transmissible *Rubus* diseases. Mild streak disease may reduce fruit yield and quality in some blackberry cultivars, but other infected cultivars may remain symptomless. Definitive data that link these graft-transmissible streak disease agents to crop damage are lacking. Mild and severe strains of the streak agent have been reported, but the possibility that additional viruses may be present, giving rise to severe streak, has not been studied. There are reports documenting the spread of mild streak in black raspberry fields and implicating wild *Rubus* plants as reservoirs, but no vectors have been reported. The use of certified black raspberry planting stock propagated from plants known to be free of black raspberry streak and the elimination of wild *Rubus* near black raspberry fields in the northeastern United States are recommended control measures.

Boysenberry Decline

Recently a disease called boysenberry decline has been noted in boysenberry in New Zealand; it causes compact rosettes of proliferated lateral shoots (Plate 78), flowers with crinkled petals, stubby stamens and enlarged styles and stigmas (Plate 79), and small, cone-shaped, "hairy" fruit with persistent styles (Plate 80). A mycoplasmalike organism is suspected to cause this graft-transmissible disease, but the etiology and mode of transmission of boysenberry decline are yet to be determined.

Bramble Yellow Mosaic Virus

Bramble yellow mosaic disease, caused by bramble yellow mosaic virus (BrYMV), is limited to one wild blackberry species (*Rubus rigidus* Smith) in one location in South Africa. It has no known economic importance.

Primocane leaves of *R. rigidus* infected with BrYMV develop a strong yellow mosaic early in the spring, which changes to a pattern of small, bright yellow, vein-associated spots later in the growing season (Fig. 29). Primocanes emerging later in the growing season remain symptomless. BrYMV has been successfully graft-transmitted to *R. henryi* Hemsl. & Kuntze, causing mild transient chlorotic leaf mottle, and to *Fragaria vesca* L. 'EM-1,' causing interveinal foliar chlorosis followed by dwarfing, leaf necrosis, and plant death. The virus is readily sap-transmissible to several *Chenopodium* spp., including *C. quinoa*. Other hosts susceptible to mechanical inoculation are *Gomphrena globosa* L. 'Rose' and *Nicotiana tabacum* L. 'White Burley.' The virus is filamentous, with a modal length of 730 nm.

No vector is known for BrYMV, and its natural spread has not been investigated. No controls have been developed for this virus, because of its known limited distribution in only one wild *Rubus* host.

Cherry Rasp Leaf Virus

Cherry rasp leaf virus (CRLV) naturally occurs in western North America in cherry trees and in weeds such as dandelion (*Taraxacum officinale* Wigg.). It has been found in *Rubus* only in a few red raspberries in Quebec. The economic importance of CRLV in commercial *Rubus* crops is not known. Experimentally, CRLV has been graft-transmitted to cultivated and wild *Rubus* hosts. It is readily mechanically transmitted to a number of common greenhouse herbaceous test plants, including *Chenopodium quinoa* Willd., *C. amaranticolor* Coste & Reyn., *Cucumis sativus* L., and *Phaseolus vulgaris* L.

CRLV is a single-stranded RNA virus with a bipartite genome. Its virion is isometric, 28 nm in diameter, and consists of three capsid proteins of relative molecular masses (M_r) of 26,000, 23,000, and 21,000 Da and two RNA species with

M_r of 2.56 and 1.26×10^6 Da. The virus can be purified from herbaceous hosts but is unstable and present in low amounts in sap. Antisera against CRLV have been prepared, and the *Rubus* and cherry isolates examined appear to be serologically indistinguishable.

CRLV is efficiently transmitted by the ectoparasitic American dagger root nematode, *Xiphinema americanum* Cobb (sensu lato). It is also seedborne in some weeds and has been recovered from cherry pollen. No information is available about the epidemiology or control of CRLV in *Rubus*.

Raspberry Yellow Spot

A graft-transmissible disorder of wild and cultivated red raspberries has been reported in Poland and is believed to be distinct from other known *Rubus* diseases. The main symptom of the disease, called raspberry yellow spot, is the development of foliar yellow spots of irregular size, shape, and distribution in the cultivar Malling Promise. These leaf spots are most conspicuous early in the growing season. Leaf curling, stunting, and plant dwarfing may occur. Raspberry yellow spot disease is transmitted by grafting and by the large raspberry aphid (*Amphorophora idaei* Börner). Wild red raspberry is suspected to be a reservoir host. The causal agent is presumed to be a virus, but direct evidence for this is lacking.

Rubus Chinese Seedborne Virus

A single, symptomless plant of an unidentified *Rubus* species grown in England from seed from the People's Republic of China was found to contain a 30-nm isometric, sap-transmissible virus distantly related to strawberry latent ringspot virus. This virus, Rubus Chinese seedborne virus (RCSV), caused chlorotic and necrotic local lesions, followed by systemic necrotic flecking, chlorosis, and apical distortion on mechanically inoculated *Chenopodium quinoa* Willd. and chlorotic local lesions and systemic chlorotic flecks and chlorosis on cucumber. The virus was readily purified in good yields from mechanically inoculated *C. quinoa* and *Nicotiana clevelandii* A. Gray to produce empty particles and infective particles having two coat protein subunits (M_r 47,000 and 25,200 Da) and two single-stranded RNA species, one of undetermined high molecular weight and the other of M_r 1.4×10^6 Da. The economic importance and distribution of this virus are unknown, but this report underscores the need to monitor the virus content of *Rubus* seeds received from foreign countries.

In the Himalayan foothills in India, a few plants of *R. ellipticus* Smith were found with chlorotic foliar rings. Symptomless *R. ellipticus* seedlings developed similar symptoms when approach-grafted to infected plants or planted in soil from roots of naturally infected plants. The identity of the virus presumed to be associated with these transmissions is unknown.

Seedborne dsRNA in Wild *Rubus*

When double-stranded ribonucleic acid (dsRNA) in the relative molecular mass (M_r) range of $1-5 \times 10^6$ Da is extracted from a plant, it is often considered to be the replicative form of one or more plant viruses or, alternatively, dsRNA virus. In British Columbia, some symptomless wild plants of *Rubus strigosus* Michx. and *R. leucodermis* Douglas ex Torr. & A. Gray were found to have dsRNA bands of M_r 3.4 and 1.7×10^6 Da. Many seedlings from *R. leucodermis* plants with these dsRNA band patterns also had the same band patterns. An alternate band pattern with lower M_r values found in other symptomless *R. leucodermis* sources was also seed-transmitted. None of these band patterns were found in cultivated red raspberry, but dsRNA bands of M_r 12 and 6×10^6 Da were often found in such cultivars that were free of known viruses. No viruslike particles were found in thin-section electron microscopy in any of the *R. strigosus* or *R. leucodermis* plants examined. A complimentary DNA probe was prepared from *R. leucodermis* dsRNA (M_r 3.4×10^6 Da). It hybridized with genomic DNA of *R. leucodermis* and *R. strigosus*. The origin and role of dsRNAs of M_r $1-3.4 \times 10^6$ Da in wild *R. leucodermis* and *R. strigosus* are unknown. These data do serve to warn against automatically attributing dsRNA bands in this M_r range in *Rubus* to the presence of plant viruses.

Thimbleberry Ringspot Virus

Thimbleberry ringspot virus (ThRSV) was found in one location in wild thimbleberry (*Rubus parviflorus* Nutt.) in Vancouver, British Columbia. This aphid-borne virus is graft-transmissible to red and black raspberry and to the indicator plant *R. henryi* Hemsl. & Kuntze. It is not considered to be a present or potential economic problem.

On thimbleberry, the only natural host, leaves of infected plants may have pronounced chlorotic ring spots and line patterns (Plate 81) involving large areas of the leaf, or symptoms may be restricted to small patches of chlorotic veins.

ThRSV is aphid- and graft-transmissible but not mechanically transmissible. Isometric viruslike particles 25 nm in

Fig. 29. Leaf of *Rubus rigidus* naturally infected with Rubus yellow mosaic disease, showing vein-associated small yellow spots coalesced into line patterns. (Reprinted, by permission, from Engelbrecht, 1987)

diameter have been seen in electron micrographs of infected leaf mesophyll cells. The relationship of ThRSV to other viruses is not known.

The virus is inefficiently transmitted in a semipersistent manner by the three aphids species that colonize thimbleberry in British Columbia (*Illinoia maxima* (Mason), *I. davidsonii* (Mason), and *Amphorophora parviflorii* Hill). Since ThRSV does not appear to spread to commercial *Rubus*, no control measures have been devised.

Tobacco Necrosis Virus

Tobacco necrosis virus is transmitted by motile spores of the chytridiaceous fungus *Olpidium brassicae* (Woronin) P. A. Dang. It occurs in the roots of a very large number of flowering plant species. It was reported once from the roots of field-grown red raspberry in Scotland but is probably often symptomless and undetected in many *Rubus* crops. No economic loss has been associated with TNV in *Rubus*, and no control for it in *Rubus* has been proposed.

Tobacco Rattle Virus

Tobacco rattle virus is transmitted by several species of the stubby-root nematode *Trichodorus*. It has a wide host range in flowering plant species, including *Rubus*, where it has occasionally been isolated in Scotland from raspberry roots. Infected plants are symptomless, and no controls have been devised for the virus in *Rubus*.

Tobacco Ringspot Virus

Tobacco ringspot virus (TobRSV) has been reported once from cultivated blackberry in British Columbia. It has also been reported in wild blackberries in North Carolina. This virus has a number of important economic and weed hosts outside the genus *Rubus*.

Wild blackberries in North Carolina infected with TobRSV had stunted, distorted foliage that exhibited faint to severe ring spots, chlorotic line patterns, mottling, and mosaic (Plate 82).

Some of the herbaceous plants commonly used for sap inoculation of TobRSV are *Cucumis sativus* L. (cucumber), *Nicotiana tabacum* L., and *Chenopodium quinoa* Willd.

TobRSV is the type member of the nepovirus group (nematode-borne polyhedral viruses). It is a virus with a bipartite genome. The isometric virions are about 28 nm in diameter, and the coat protein subunit has a relative molecular mass (M_r) of 55,000 Da. There are three classes of virions that have sedimentation coefficients of 53, 91, and 126 and differ in their RNA content. The two single-stranded RNAs have M_r values of 2.4 and 1.4×10^6 Da. Both RNAs are needed for infection. Natural transmission is by the dagger nematode *Xiphinema americanum* Cobb (sensu lato). Antisera specific for TobRSV are available and can be used for rapid, accurate detection of this virus with enzyme-linked immunosorbent assay.

There are no studies on the spread or control of TobRSV in *Rubus*, because of its rare occurrence.

Selected References

Barbara, D. J., Ashby, S. C., and McNamara, D. G. 1985. Host range, purification and some properties of Rubus Chinese seed-borne virus. Ann. Appl. Biol. 107:45-55.

Basak, W. 1974. Yellow spot—A virus disease of raspberry. Bull. Acad. Pol. Sci. Ser. Sci. Biol. Cl. 5. 22:47-51.

Bos, L. 1970. Bean yellow mosaic virus. Descriptions of Plant Viruses, No. 40. Commonwealth Mycological Institute and Association of Applied Biologists, Kew, Surrey, England. 4 pp.

Cadman, C. H. 1961. Raspberry virus and virus diseases in Britain. Hortic. Res. 1:47-61.

Converse, R. H. 1970. Black raspberry streak. Pages 155-157 in: Virus Diseases of Small Fruits and Grapevines. N. W. Frazier, ed. University of California, Division of Agricultural Sciences, Berkeley.

Converse, R. H. 1986. Sterility disorder in 'Darrow' blackberry. HortScience 21:1441-1443.

Converse, R. H., and Lister, R. M. 1969. The occurrence and some properties of black raspberry latent virus. Phytopathology 59:325-369.

Engelbrecht, D. J. 1963. Sap-transmission of viruses from pear, grapevines and bramble to herbaceous hosts. S. Afr. J. Lab. Clin. Med. 9:140-141.

Engelbrecht, D. J. 1976. Some properties of a yellow mosaic virus isolated from bramble. Acta Hortic. 66:79-83.

Engelbrecht, D. J. 1987. Bramble yellow mosaic. Pages 243-244 in: Virus Diseases of Small Fruits. R. H. Converse, ed. U.S. Dep. Agric. Agric. Handb. 631.

Engelbrecht, D. J., and van der Walt, W. J. K. 1974. Host reaction and some properties of a virus causing yellow mosaic in wild bramble (*Rubus* sp.) Phytophylactica 6:311-313.

Hansen, A. J., Nyland, G., McElroy, F. D., and Stace-Smith, R. 1974. Origin, cause, host range and spread of cherry rasp leaf disease in North America. Phytopathology 64:721-727.

Hemphill, D. D. 1970. Sterility in *Rubus*. Page 159 in: Virus Diseases of Small Fruits and Grapevines. N. W. Frazier, ed. University of California, Division of Agricultural Sciences, Berkeley.

Horn, N. L., and Woods, M. W. 1949. Transmission of the mild streak virus of black raspberry. Phytopathology 39:377-385.

Jeffers, W. F., and Woods, M. W. 1948. Field studies on spread of the mild streak disease of black raspberries. Phytopathology 38:222-226.

Jones, A. T., Mayo, M. A., and Henderson, S. J. 1985. Biological and biochemical properties of an isolate of cherry rasp leaf virus from red raspberry. Ann. Appl. Biol. 106:101-110.

Lister, R. M., and Converse, R. H. 1972. Black raspberry latent virus. Descriptions of Plant Viruses, No. 106. Commonwealth Mycological Institute and Association of Applied Biologists, Kew, Surrey, England. 4 pp.

Nyland, G., Lownsbery, B. F., Lowe, S. K., and Mitchell, J. F. 1969. The transmission of cherry rasp leaf virus by *Xiphinema americanum*. Phytopathology 59:1111-1112.

Provvidenti, R., and Granett, A. L. 1974. Sweet violet, a natural host of bean yellow mosaic virus. Plant Dis. Rep. 58:155-156.

Rush, M. C., and Gooding, G. V., Jr. 1970. The occurrence of tobacco ringspot virus strains and tomato ringspot virus in hosts indigenous to North Carolina. Phytopathology 60:1756-1760.

Stace-Smith, R. 1958. Studies on *Rubus* virus diseases in British Columbia. 5. Thimbleberry ring spot. Can. J. Bot. 36:385-388.

Stace-Smith, R. 1985. Tobacco ringspot virus. Descriptions of Plant Viruses, No. 309. Commonwealth Mycological Institute and Association of Applied Biologists, Kew, Surrey, England. 6 pp.

Stace-Smith, R. 1987a. Thimbleberry ringspot. Pages 192-194 in: Virus Diseases of Small Fruits. R. H. Converse, ed. U.S. Dep. Agric. Agric. Handb. 631.

Stace-Smith, R. 1987b. Tobacco ringspot virus in *Rubus*. Pages 227-228 in: Virus Diseases of Small Fruits. R. H. Converse, ed. U.S. Dep. Agric. Agric. Handb. 631.

Stace-Smith, R., and Hansen, A. J. 1974. Occurrence of tobacco ringspot virus in sweet cherry. Can. J. Bot. 52:1647-1651.

Stace-Smith, R., and Hansen, A. J. 1976. Cherry rasp leaf virus. Descriptions of Plant Viruses, No. 159. Commonwealth Mycological Institute and Association of Applied Biologists, Kew, Surrey, England. 4 pp.

Stace-Smith, R., and Hansen, A. J. 1976. Some properties of cherry rasp leaf virus. Acta Hortic. 67:193-197.

Stace-Smith, R., and Martin, R. R. 1988. Occurrence of seed-transmissible double-stranded RNA in native red and black raspberry. Acta Hortic. 236:13-20.

(Prepared by R. H. Converse)

Nematode Parasites

When the U.S. Department of Agriculture published its Agriculture Handbook 310 in 1905, little information on the importance of nematodes on *Rubus* spp. was available from North America, but in Europe nematode-transmitted viruses were an established and well-documented problem. Earliest reports from North America appeared in the mid-1930s from Canada, where *Pratylenchus* spp. were associated with declining red raspberries. In the 1950s, several reports were published on nematodes associated with poorly growing red raspberries and blackberries in the United States and Canada.

While some progress has been made in evaluating the threshold densities and management of certain nematode species, more information is needed regarding the importance of some species and the pathogenic and virus-vector capabilities of others.

Root-Lesion Nematodes

Root-lesion nematodes (*Pratylenchus* spp.) are widely associated with *Rubus* spp. throughout the world. Six *Pratylenchus* spp. have been found in unthrifty raspberry plantations, but only *P. penetrans* (Cobb) Filipjev & Schuurmans, Steckhoven appears to be of consequence in the major *Rubus* production areas of the world. *P. vulnus* Allen & Jensen has been associated with red raspberry decline in Utah, and *P. crenatus* Loof is of minor importance in Europe.

Symptoms

Root symptoms of plants lightly infected or in the early stages of damage are difficult to detect; they appear only as small, elongate discolorations or lesions on the new roots (thus the common name *root-lesion nematode*). As populations increase and more damage occurs, feeder roots die, which stimulates a proliferation of fine roots and results in a witches'-broom symptom. In the final stages all feeder roots die, leaving only the large-diameter roots, which have little capacity for nutrient and water uptake. These are frequently invaded by secondary fungi, which contribute to decay.

Above the ground, the cane growth parallels the root decline. Slight stunting is the only aboveground symptom of lightly infected plants or those in the early stages of decline. Without comparative nematode-free plants, damage is usually imperceptible. Reduced cane number, diameter, and general plant vigor (along with some dieback) mark the slow decline (Plate 83). Increased susceptibility to winter injury, resulting in dead buds, is a frequent secondary symptom of nematode-stressed plants.

The rate of decline depends upon the nematode population density but usually occurs over a 3- to 4-year period. Raspberry plantations of susceptible varieties with populations at or just below detectable levels at the time of planting in loam soils generally begin to decline in years 6–8, with die-out occurring 2–3 years after the onset of decline (Plate 84). The rate of decline may be increased by the presence of other stresses, such as drought, disease, insects, and infertility.

Causal Organism

P. penetrans is but one of the species of nematodes referred to as root-lesion nematodes, which elicit lesions on roots. Revision of the genus *Pratylenchus* in 1953 suggests that earlier reports of *P. pratensis* (de Man) Filipjev associated with declining red raspberry are in fact those of *P. penetrans*. *P. pratensis* has not been found causing damage to *Rubus* spp., but it is occasionally associated with those crops.

Life Cycle and Epidemiology

Root-lesion nematodes are migratory endoparasites, spending a portion of their lives in the soil and in root tissues. As they migrate back and forth, they deposit eggs singly in either soil or roots. Females lay an average of one egg per day. The first-stage larva develops within the egg and passes through one molt inside the egg before emerging as a second-stage larva. Each larval stage must feed before molt can occur. The life cycle is completed in about 30–100 days, depending on soil temperature.

All larval and adult stages are capable of invading roots. They feed on cells for a period before cutting a slit in the cell wall with their stylet. This process is repeated as the nematode moves from cell to cell throughout the root cortex. Cell and root destruction results from direct nematode feeding and invasion by secondary fungi. Host physiology may also be altered, increasing susceptibility to other pathogens. Many root-lesion nematode species parthenogenetically reproduce, and the presence or absence of males in a population is frequently used as one of the diagnostic characters of the species.

Root-lesion nematodes in red raspberry plantations are usually active throughout the year, but activity significantly decreases during freezing or dry soil conditions. During periods of cold or drought, nematodes move deeper in the soil profile or into roots, where they are protected. Active *P. penetrans* have been recovered from red raspberry soil and roots under 20 cm of snow and in soil frozen to a depth of 15 cm.

The most accurate assays of nematode populations are carried out in the fall, when densities are at their greatest and more indicative of the damage potential. Population-damage assessments at other times of the year must be extrapolations of the fall populations.

In established red raspberry plantings with areas of poor growth, a comparison of nematode populations in the soil and roots from areas with good growth and those from areas with poor growth best determines the impact of nematodes on the crop. When population densities in poor but not dead areas are 10 or more times that of good areas, they are considered to be involved in the cause of the problem and require treatment.

Control

Exclusion. *P. penetrans* is widespread and endemic in most red raspberry-growing regions worldwide. New fields, even from newly cleared land, should be surveyed prior to planting. Only nematode-free stock should be planted in fields found free of nematodes or following fumigation. Nematodes are transported in soil and water, primarily by humans and farm machinery. Once established, nematode infestations are permanent.

Pre-planting fumigation. All nursery stock should be grown in fumigated soil. Proper planning to ensure adequate soil preparation and conditions at the time of treatment are the keys to successful fumigation. Application of a systemic herbicide to the previous crop kills the roots and forces the root-lesion nematodes into the soil, where they are more easily controlled. This should be done far enough in advance of fumigation to allow the plowed-down crop to decompose. Large amounts of organic matter interfere with the effectiveness of fumigants. Deep ripping (0.8–1.5 m) of the soil at a 1-m spacing in two perpendicular directions enables the fumigant to reach nematodes deep in the soil. The soil should be worked into a seedbed condition, with the breaking up of large clods that might protect nematodes from the fumigant.

Chemical manufacturers' directions should be followed for any soil-applied pesticide to determine rates for various soil types, soil temperatures and moistures, and treatment and aeration times.

A mixture of methyl bromide and chloropicrin (67:33) at 280–392 kg/ha under continuous polyethylene (2-mil or thicker) tarp has been shown to be cost-effective in controlling nematodes, soilborne fungi, insects, and weeds. Planting disease-free stock, combined with soil monitoring, allows 2 years' production from a single fumigation, thus reducing the per annum cost of treatment.

For many growers whose primary concern is fruit production, the high cost of fumigation with a methyl bromide and chloropicrin mixture is not always justified. The only preplant fumigant nematicide currently registered for use in *Rubus* plantations in the United States is 1,3-dichloropropene (1,3-D) at 234–280 L/ha. Methyl bromide at 224–280 kg/ha without tarp has been demonstrated to be cost-effective in controlling nematodes under certain circumstances. Where other soilborne diseases and pests are present, a more general biocide (such as methyl bromide and chloropicrin under tarp, methyl isothiocyanate, metam-sodium, and dazomet) may be necessary.

Population densities exceeding 500 nematodes per 500 cm^3 of soil affect stand establishment, first-year vegetative growth, and yield. This is called the economic threshold density. Nematode populations above this level cause economic damage. Fields in this category should be fumigated prior to planting. Below this density fumigation is optional, provided an adequate nematicide is available for use on established plantings.

Treatment of established plantings. Declining yield and plant growth indicates that root and soil samples should be taken as described above to determine if nematodes are present. Use of a postplanting nematicide should be based on nematode assays. The population density dictates the nematicide rate to be applied.

Currently, the only nematicide registered for use on established red raspberry in the United States is fenamiphos. It is not registered for use on blackberry. Studies in the northwestern United States have shown that root-lesion nematode densities of 1,000 and 4,000 per 500 cm^3 of soil require treatment rates of 6.7–13.5 kg a.i. per hectare of fenamiphos, respectively. These treatments are effective in reducing populations and increasing plant growth and yield. It may not be applied later than 6 months prior to harvest. Fenamiphos, a nonfumigant nematicide, does not kill nematodes immediately but interferes with feeding and reproduction. Because populations decrease slowly, sampling to evaluate the efficacy of treatment should be delayed 3–6 months following treatment.

It may take up to 3 years to detect the yield responses that result from decreased nematode populations. After treatment, increased root production occurs, followed by increased primocane production in the second year and fruit production on these canes in the third year. For this reason, nematode densities should be evaluated annually in order to determine where problem areas exist. Annual evaluations also allow growers to initiate treatment before decline symptoms develop.

Selected References

Golden, A. M., and Converse, R. H. 1965. Nematodes on raspberries in the Eastern United States. Plant Dis. Rep. 49:987-991.

Jennings, D. L. 1988. Diseases caused by nematodes. Pages 139-142 in: Raspberries and Blackberries: Their Breeding, Diseases and Growth. Academic Press, New York.

McElroy, F. D. 1973. Control of *Pratylenchus penetrans* in raspberry root cuttings. Plant Dis. Rep. 57:492-495.

McElroy, F. D. 1977. Effect of two nematode species on establishment, growth and yield of raspberry. Plant Dis. Rep. 61:277-279.

McElroy, F. D. 1972. Nematodes in tree fruits and small fruits. Pages 335-376 in: Economic Nematology. J. M. Webster, ed. Academic Press, New York.

McElroy, F. D. Nematode management in brambles—A plant health care program. J. Nematol. In press.

Trudgill, D. L. 1983. The effect of nitrogen and of controlling *Pratylenchus penetrans* with nematicides on the growth of raspberry (*Rubus idaeus* L.) Crop Res. 23:103-112.

(Prepared by F. D. McElroy)

Dagger and Needle Nematodes

Three species of dagger nematodes (*Xiphinema* spp.) have been associated with poor growth of *Rubus* spp. *X. bakeri* Williams has the greatest direct impact on plant growth by root destruction. *X. americanum* Cobb and *X. diversicaudatum* (Micoletzky) Thorne have the greatest impact by serving as vectors for tomato ringspot nepovirus (TomRSV) and the Arabis mosaic/strawberry latent ringspot nepovirus (AMV/SLRV), respectively. Nepoviruses are nematode-transmitted, polyhedral viruses. Virus particles of this type are selectively retained in the esophageal region of the vector nematodes during feeding on virus-infected plants and may be retained in an infective state from several weeks to a year or more. However, the virus is not retained through molting or passed to the eggs. During the feeding of viruliferous nematodes, virus particles are released into the plant cell, initiating virus infection.

X. americanum causes little direct damage to *Rubus*, but *X. diversicaudatum* causes some root galling and decline in red raspberry. The former is found widely distributed in North America. The latter is widely distributed in Europe, but it has been reported only in greenhouses on Long Island in New York. *X. bakeri* is not known to transmit any virus in North America; however, it does cause significant direct damage to red raspberry roots.

Needle nematodes (*Longidorus* spp.) appear to cause little direct damage, but substantial damage is caused by the plant viruses they transmit. Raspberry ringspot nepovirus (RRV) is transmitted by *L. elongatus* de Man and *L. macrosoma* Hooper. Tomato black ring nepovirus (TBRV) is transmitted by *L. elongatus* and *L. attenuatus* Hooper, and peach rosette virus is transmitted by *L. diadecturus* Eveleigh & Allen. *L. elongatus* is found in North America, but not in association with RRV, and apparently does not cause any direct damage to *Rubus* spp. Neither *L. macrosoma* nor *L. attenuatus* or associated viruses are found in North America; however, *L. diadecturus* and its associated viruses are endemic in North America but thus far have not been associated with any damage to *Rubus* spp.

Symptoms

Extensive feeding of *X. bakeri* and *X. diversicaudatum* on raspberry and blackberry roots causes gall formation and irreversible cessation of root elongation. Feeding by *X. bakeri* causes swelling and "fishhook" curling of the root tip, which may be used as a field diagnostic aid (Plate 85). This symptom results from nematodes feeding on only one side of the root, which stops cell development in that region, while normal cell development continues on the other side, thus causing the "fishhook" curvature of the root. The development of lateral roots is also prevented on other portions of the root.

Stunting of the root system also results in poor cane growth. As few as 100 nematodes per 500 cm^3 of soil can reduce root and top growth by 40–50%. Decline similar to that described for plants attacked by *Pratylenchus penetrans* also results from attack by *X. bakeri*. *X. americanum* causes no direct damage to *Rubus* spp., but it is the vector for TomRSV, which results in a gradual decline over several years. In addition, spring leaf-out of TomRSV-infected red raspberry plants is retarded by 2–3 weeks; virus-free plants suffer no retardation. The most important consequence of infection is that the virus causes crumbly fruit and reduces yields. In severe situations, damage from virus infection necessitates removal of the unproductive planting. (These disease symptoms, as well as those caused by other nematode-transmitted viruses, are described in more detail in the virus section.)

Causal Organism

X. bakeri was first described as a new species from specimens associated with raspberry plants at Hatzic, British Columbia, in 1961.

Currently, the identification of nematodes from the *X. americanum* group is controversial; 32 nomial species have been described, and it is likely that several of these are variants of species. In 1965, *X. americanum* was reported to be associated with, but not directly affecting, raspberry. It was shown to transmit TomRSV to a number of crops, including *Rubus* spp. In 1982, Wortowicz et al confirmed that *X. americanum* and *X. rivesi* Dalmasso frequently occurred together in the same soils in the eastern United States. The authors also identified a population of *Xiphinema* from *Rubus* spp. in Vermont as *X. rivesi*. Ebsary et al in 1984 affirmed the presence of *X. americanum* and *X. rivesi* in eastern Canada and described *X. occidium* Ebsary, Potter & Allen as a new species in British Columbia soils. However, Lamberti and Golden suggested in 1984 that *X. occidium* was, in fact, *X. thornei* Lamberti & Golden. They also indicated that *X. americanum* sensu lato does not occur in the western United States, suggesting that those species were probably *X. rivesi*, *X. thornei*, *X. pachtaicum* (Tulaganov) Kirjanova, or *X. califoricum* Lamberti & Beleve-Zachero. In 1989, Ebsary et al reassessed *X. occidium*, referring to populations from Alberta, Saskatchewan, and Manitoba as *X. occidium* and populations from British Columbia as *X. bricolensis* Ebsary, Vrain & Graham, which represented a new species. A second species, *X. pacificum* Ebsary, Vrain & Graham, was also identified from British Columbia.

The taxonomy of the *X. americanum* group remains controversial, and differential transmissibility of variants of TomRSV by populations of the nematode have been reported. Studies are currently in progress to sort out this confusion. Until this work is completed and agreement on species designation is reached, it is probably judicious to consider all *X. americanum*-group populations in North America as possible vectors of TomRSV.

Life Cycle and Epidemiology

Needle and dagger nematodes are similar to the lesion nematode in life cycle and development. They differ slightly from the lesion nematode, because they hatch from eggs as first-stage rather than second-stage juveniles. Completion of the life cycle of needle and dagger nematodes from egg to egg is dependent on soil temperatures but may take as little as 30 days. Adults may live for 2–3 years.

Dagger and needle nematodes are called *ectoparasites*, since they spend their entire lives outside the root. Only their very long stylet, a hypodermic-needle-like feeding structure, penetrates deep into the root tissue to feed and transmit viruses. Males are rare or not known for many species, since reproduction is mostly parthenogenetic. The *X. americanum* group reproduces parthenogenetically and thus may be considered a conglomerate of monosexual forms.

Like the root-lesion nematode, many dagger and needle nematodes are native to *Rubus* production areas. Nepoviruses are also present in many native weeds, from which they may be acquired and transmitted by *Longidorus* and *Xiphinema*. It is extremely important to sample potential berry fields, even if they are being developed from native vegetation, to ensure that the virus and vector are absent. It is unlikely that these nematodes will be introduced to sites in planting material, but the nepoviruses can be readily introduced by this method. It is therefore imperative that only virus-free planting stock be used in establishing new fields.

Control

Strategies for the control of dagger and needle nematodes are similar to those for root-lesion nematodes. Exclusion techniques also are the same, except that transport of dagger and needle nematodes in bare root planting stock is less likely, since they are ectoparasites. However, it is important to ensure they are not introduced into a new area, since they may be carrying their associated nepoviruses, which would result in a permanent field infestation of both organisms.

In Britain several red raspberry cultivars were bred with resistance to some of the European nematode-transmitted nepoviruses. The original resistance screening was done by means of graft transmission with virus-infected material. However, Jones et al in 1989 observed that several of these cultivars were susceptible to virus infection when planted in sites where viruliferous nematodes were present. Conversely, some susceptible cultivars were found to be resistant to virus infection at these sites.

Standard preplant fumigation treatments for nematode control may be used, but economic thresholds for treatment are considerably lower. The presence of a single vector, especially if it is viruliferous, necessitates preplant fumigation. Neither the virus nor the vector can be controlled by current postplanting techniques.

Fields in areas known to have both the virus and vector should be sampled 2 years prior to planting. If the vector is detected, growers should plant a shallow-rooted grass crop for 1–2 years. This shallow-rooted crop brings nematodes to the upper part of the soil profile, where they are more easily controlled. The use of a nonhost crop also reduces virus inoculum. Subsequent generations of nematodes will then be virus-free, even if they survive soil fumigation. Excellent weed control is essential, since nepoviruses have a wide host range on nongrass crops and weeds and are symptomlessly seedborne in many weed hosts.

For nonvector nematodes, such as *X. bakeri*, economic thresholds for treatment are intermediate between *P. penetrans* and *X. americanum*. Levels of *X. bakeri* above 50 nematodes per 500 cm^3 of soil require treatment to prevent adverse effects on stand establishment. Populations of *X. bakeri* should be reduced to extremely low densities prior to planting, because there presently are no postplanting control measures. Fortunately, this species can be fairly easily controlled by bare soil fallow, with repeated turning and drying of the soil through the summer months.

Selected References

Dojatowicz, M. R., Golden, A. M., Forer, L. B., and Stouffer, R. F. 1982. Morphological comparisons between *Xiphinema rivesi* Dalmasso and *Xiphinema americanum* Cobb populations from the eastern United States. J. Nematol. 14:511-516.

Ebsary, B. A., Potter, J. W., and Allen, W. R. 1984. Redescription and distribution of *Xiphinema rivesi* Dalmasso, 1969 and *Xiphinema americanum* Cobb, 1913 in Canada with a description of *Xiphinema occidium* n.sp. (Nematoda: Longidoridae). Can. J. Zool. 62:1696-1702.

Ebsary, B. A., Vrain, T. C., and Graham, M. B. 1989. Two new species of *Xiphinema* (Nematoda: Longidoridae) from British Columbia vineyards. Can. J. Zool. 67:801-804.

Jones, A. T., McElroy, F. D., and Brown, D. J. F. 1981. Tests for transmission of cherry leaf roll virus using *Longidorus, Paralongidorus,* and *Xiphinema* nematodes. Ann. Appl. Biol. 99:143-150.

Jones A. T., Mitchell, M. J., and Brown, D. J. F. 1989. Infectibility of some new raspberry cultivars with arabis mosaic and raspberry ringspot virus and further evidence for variation in British isolates of these two nepoviruses. Ann. Appl. Biol. 115:57-69.

Lamberti, F. and Beleve-Zachero, T. 1979. Studies on *Xiphinema americanum* sensu lato with descriptions of 15 new species (Nematoda: Longidoridae). Nematol. Mediterr. 7:51-106.

McElroy, F. D. 1972. Studies on the host range of *Xiphinema bakeri* and its pathogenicity to raspberry. J. Nematol. 4:16-22.

Valdez, R. B. 1972. Transmission of raspberry ringspot virus by *Longidorus caespiticola, L. leptocephalus* and *Xiphinema diversicaudatum* and arabis mosaic virus by *L. caespiticola* and *X. diversicaudatum*. Ann. Appl. Biol. 71:229-234.

Williams, T. D. 1961. *Xiphinema bakeri* n.sp (Nematoda: Longidoridae) from the Fraser River Valley, British Columbia, Canada. Can. J. Zool. 39:407-412.

(Prepared by F. D. McElroy)

Other Nematodes

Several other genera of nematodes have been associated with raspberry and blackberry from earliest records, but none have been confirmed as pathogens. *Criconema, Criconemoides, Hemicycliophora,* and *Meloidogyne* are among these genera. The raspberry cultivars Canby and Newburgh have been shown to be hosts for *M. hapla* Chitwood, but the only damage appears to be an increased incidence of the crown gall symptom caused by *Agrobacterium tumefaciens* (Smith & Townsend) Conn.

Selected References

Chamberlain, G. C., and Putnam, W. L. 1955. Diseases and insect pests of the raspberry in Canada. Can. Dep. Agric. Publ. 880 (rev.).

Christie, J. R., and Taylor, A. L. 1958 Controlling nematodes in the home garden. U.S. Dep. Agric. Bull. 2048.

Griffin, J. D., Anderson, J. L., and Jorgenson, E. C. 1968. Interaction of *Meloidogyne hapla* and *Agrobacterium tumefaciens* in relation to raspberry cultivars. Plant Dis. Rep. 52:492-493.

(Prepared by F. D. McElroy)

Part II. Arthropod Pests

Insects and related arthropods, such as mites, directly attack and cause damage on all parts of the *Rubus* plant, including roots, crown, canes, foliage, and fruit. In addition, insects cause extensive indirect damage by vectoring destructive plant viruses. A relatively new concern is the contamination of mechanically harvested fruit with insects and their body parts.

Insect and mite pests of raspberry and blackberry often provide major constraints to commercial production. The number and intensity of arthropod pests vary from one location to another. However, regardless of where *Rubus* spp. are grown, a few economically important pests are generally present. Proper identification of these pests is critical to selecting appropriate control strategies, be they cultural, biological, mechanical, or chemical. Often pests are not present when damage is first observed. Thus, it may be necessary to identify a pest by the damage it has caused. Improper pest identification could be a costly mistake for commercial growers, particularly when they are selecting insecticides for chemical control.

This section provides pertinent information about the life history, identification, and characteristic damage caused by the more common and destructive pests of *Rubus*. General control strategies are also discussed. Because of rapidly changing regulations and restrictions governing pesticide usage, specific insecticide recommendations are not provided. Nonchemical control strategies are discussed where appropriate.

Insects That Damage Roots and Crowns

Raspberry Crown Borer

The adult of the raspberry crown borer, *Pennisetia marginata* (Harris), is a clearwing moth that resembles a yellowjacket in both markings and habits (Plate 86). It is one of the most damaging pests of *Rubus* crops in North America. Its range extends from the Canadian Pacific Northwest to Florida in the Southeast. The movement of this clearwing moth into a field is so gradual that it goes practically unnoticed. Even after a field is heavily infested, and cane vigor is reduced, it is difficult to detect the insect, because of the cryptic nature of the larval stage. In Europe, *P. hylaeiformis* (Laspeyes) and *P. bohemica* (Králíček & Povolný) feed on the crowns of various *Rubus* spp., causing similar damage.

Life History and Description

The adult is in the same family (Sesiidae) as the peach tree borer, lesser peach tree borer, and strawberry crown borer. It has a black body with four yellow transverse stripes. Unlike most other bramble pests, it lays its eggs individually near the edge of the underside of leaflets in late summer. Females land on foliage during the day and lay reddish brown eggs. After hatching, the larvae (Plate 87) migrate to the base of the cane. The larvae begin their 2-year life cycle by forming a blisterlike hibernaculum just below soil level at the base of a stem. In the spring, the larvae bore numerous galleries, and by the middle of the second summer the crown may be extensively damaged. At midsummer pupation begins, and by late August the clearwing moths begin to emerge. After mating, a female lays about 140 eggs, and a new cycle begins.

Young larvae are small and white, with a light brown head. By the second winter they are 1.3–2 cm long. By the following July they are mature, 2.5–3.3 cm long, and they soon pupate inside the base of the plant. The pupae are about 2 cm long and reddish brown in color.

Damage

Aboveground symptoms are visible only to the trained eye. Loss of vigor and spindly canes can be confused with symptoms of several diseases. The most obvious symptom is individual canes that are withered, wilted, and bent over with dying or dead foliage. This symptom commonly occurs when fruit is about half-grown. Another common symptom is the lodging of affected canes. Positive identification is obtained by putting on gloves and giving a sharp tug to damaged canes. Those infested by the borer will break away easily, revealing borer damage at the point of breakage. Often a larva is found inside the portion of the cane that breaks off.

Injury to the crown can be observed only if the plant is dug up and the crown examined. Signs of crown borer infestation are girdled roots and crowns and cavities burrowed into the exterior of the crowns. It is usually necessary to cut open the crown in order to find the larva or pupa.

Control

The removal of alternate hosts, such as wild brambles (particularly wild blackberries), from the area helps reduce crown borer populations. Insecticides applied as drenches over the row in early spring are used to kill a portion of the larvae. However, it is very difficult to obtain effective control with insecticides, because soil moisture, texture, and organic matter may greatly alter efficacy. It may be necessary to treat badly infested fields several years in a row to get adequate control.

Selected References

Raine, J. 1962. Life history and behavior of the raspberry crown borer *Bembecia marginata* (Harr.) (Lepidoptera: Aegeriidae). Can. Entomol. 94(11):1216-1222.

Raine, J. 1964. Effect of raspberry crown borer, *Bembecia marginata*, on yield of loganberries. Can. J. Plant Sci. 44:75-77.

Schaefers, G. A. 1974. Raspberry crown borer: Control in relation to period of egg hatch. J. Econ. Entomol. 67:451-52.

Wylie, W. D. 1970. The raspberry crown borer, a serious pest of blackberries. Arkansas Farm Res. 19(6):6.

(Prepared by R. N. Williams)

Root Weevils

Many species of root weevils may be found on *Rubus* spp. However, the black vine weevil (Plate 88), *Otiorhynchus sulcatus* (F.), the strawberry root weevil, *O. ovatus* (L.), and the obscure root weevil, *Sciopithes obscurus* Horn, are most often encountered in North America. The black vine weevil is the most common of the three species on raspberries. In the United Kingdom, the clay-colored weevil (Plate 89), *O. singularis* (L.), is the most common.

Life History and Description

The life cycles of the three common North American species are similar. Adult weevils emerge from the soil on approximately June 1. Clay-colored weevils emerge in late April or early May. The *Otiorhynchus* weevils lay eggs on or in the soil, but *S. obscurus* oviposits along the margin of a leaf and folds the leaf over the eggs. After the eggs hatch, the larvae burrow into the soil, where they feed on plant roots. Feeding continues throughout the summer and into winter. Pupation occurs in late spring in most species. Clay-colored weevils are the exception, pupating in late summer. Adult *Otiorhynchus* weevils often survive the first winter after adult eclosion and oviposit from early May to early September of the next year.

The root weevils have very wide host ranges. Perennial weeds, including grasses, within or along the edges of fields may serve as sources of weevils to infest berry fields.

Damage

In general, relatively little damage from root weevil feeding appears to occur on *Rubus*. Red raspberry seems to tolerate feeding by root weevil larvae on its roots rather well. It is rare for plants to be visibly reduced in vigor by the root feeding. Black raspberry and blackberry virtually never suffer visible damage. Adult clay-colored weevils feeding in the spring can cause considerable damage to red raspberry by chewing the expanding buds or developing fruiting laterals. In a severe attack, all the fruit-bearing buds can be damaged. In North America, perhaps the most serious problem is that adult weevils have become a serious contaminant of mechanically harvested red raspberries. The action of the mechanical harvester causes the weevils to fall off with ripe berries. They must then be manually removed from the sorting belts on the harvester or in the processing plant. This can be very difficult if large numbers of weevils are present, or if they have had time to crawl inside the hollow berries.

Control

Fields should be kept free of perennial weeds, since weevil larvae have been observed on the roots of such plants. Weeds in fence rows, adjacent fields, and wooded or brushy areas may also be sources of root weevils.

Adult root weevils are most often detected by raspberry growers among mechanically harvested berries. Weevils may be suspected in a field because of the characteristic notching of the edges of leaflets, but other insects or mechanical injury may cause similar damage. Weevil populations can also be monitored by digging in the soil around the plant crowns in late winter or early spring to search for larvae. Adults may be detected in late spring or in summer by spreading a cloth beneath the rows at night and shaking the plants to dislodge adult weevils, which fall onto the cloth. If any weevils are found, control measures should be considered, since their populations can greatly increase within 1 year.

Insecticide sprays are the only means currently available for controlling adult root weevils on raspberries. Spray efficacy is enhanced if they are applied at night, when these nocturnal insects are active.

Selected References

Garth, G. S., and Shanks, C. H., Jr. 1978. Some factors affecting infestation of strawberry fields by the black vine weevil in western Washington. J. Econ. Entomol. 71:443-448.

Gordon, S. C., and Woodford, J. A. T. 1986. The control of the clay-coloured weevil (*Otiorhynchus singularis*) (Coleoptera: Curculionidae) in eastern Scotland. Crop Res. 26:111-119.

Smith, F. F. 1932. Biology and control of the black vine weevil. U.S. Dep. Agric. Bull. 325. 45 pp.

Wilcox, J., Mote, D. C., and Childs, L. 1934. The root-weevils injurious to strawberries in Oregon. Oreg. Agric. Exp. Stn. Bull. 330. 109 pp.

(Prepared by C. H. Shanks, Jr.)

Strawberry Crown Moth

The strawberry crown moth (Plate 90), *Synanthedon bibionipennis* (Boisduval), is a serious pest of strawberries that also occurs occasionally on red and black raspberries and loganberries in the Pacific Northwest. The larvae feed on and damage the crown and root tissues of host plants as well as facilitate entry of root pathogens. Both the strawberry crown moth and the raspberry crown borer belong to the family of clearwing moths known as Sesiidae.

Life History and Description

The crown moth overwinters as a nearly mature but inactive larva. As temperatures rise in the spring, the larvae feed for a period of 2-6 weeks. Pupation occurs during May and June within frass-covered silk cocoons. Moth emergence occurs in late spring through early summer, with peak emergence in late June or early July in the Pacific Northwest. Eggs are deposited singly on the underside of leaves close to the soil surface. These eggs hatch in 10-14 days, and the newly eclosed larvae migrate to the crown and root systems of the host plants. The larvae burrow into the crown and feed on the interior portion of the main root until late October or early November, when silken cells are produced for overwintering. There is one generation per year.

The larvae and damage of the crown moth are at times confused with those of the raspberry crown borer. However, the larvae of these two species can be distinguished by observing the last row of prolegs on the abdomen. Crochets are present on the larvae of the crown moth and absent on the larvae of the crown borer. Additionally, crown borer larvae eventually tunnel well into the crowns near the bases of the canes, whereas larvae of the crown moth normally girdle the roots and lower crowns with superficial tunnels.

Damage

Plants suffer from feeding injury to the roots and crown. Injured plants are stunted and display poor vigor. Premature yellowing, senescence, and dehiscence of leaves from canes are general symptoms of strawberry crown moth infestations in *Rubus*.

Control

Certain cultural practices help to reduce crown moth infestations in *Rubus*. Light infestations may be minimized by removing and destroying infested plants. Infested strawberry fields adjacent to susceptible *Rubus* plantings that are being removed from production should be plowed down after harvest in September. This concentrates egg deposition by the crown moth on the more favored strawberry plants during the summer flight period. *Rubus* plantings should not be established next to infested strawberry or *Rubus* fields.

Pheromone lures and traps are available for detecting and monitoring male crown moth populations. These traps have been used successfully by strawberry growers in the Pacific Northwest to determine the severity and extent of crown moth infestations in strawberry. They are also effective in determining the presence of this pest in *Rubus*.

Insecticides have been effective in controlling crown moth in commercial *Rubus* fields. Also, the border rows of susceptible *Rubus* plantings that are adjacent to infested strawberry fields should be treated with insecticides. A suitably labeled insecticide drench applied in the fall or spring will generally give adequate control if adequate water is used and soil moisture content is high. When applied as a crown drench at either of these times, control of the raspberry crown borer also occurs.

Selected References

Antonelli, A. L., Shanks, C. H., Jr., and Gisher, G. C. 1988. Small fruit pests: Biology, diagnosis and management. Wash. State Univ. Coop. Ext. Serv. Ext. Bull. 1388. 20 pp.

Fisher, G. C., and Sheets, W. A. 1984. The strawberry crown moth, a pest of strawberries and caneberries. Oreg. State Univ. Ext. Circ. 1175. 4 pp.

Weinzierl, R. A., Fisher, G. C., and Calkin, J. D. 1984. Selecting and monitoring pheromone traps in insect pest management. Oreg. State Univ. Ext. Circ. 1207. 8 pp.

(Prepared by G. C. Fisher)

Insects That Damage Fruit

Scarab Beetles

Japanese Beetle

The Japanese beetle, *Popillia japonica* Newman, was first observed in North America in New Jersey in about 1916. Since then, its continuous range has extended from the central areas of South Carolina, Georgia, and Alabama to Tennessee, eastern Missouri, and Nebraska and northward into Ontario. The Japanese beetle has also been found in California and Oregon. It is associated with grassy areas, where larvae feed on root hairs of grasses, and adults lay eggs in and emerge from turf.

Life History and Description

The adult beetle (Plate 91) is about 13 mm long and has a shiny metallic green head, thorax, and abdomen and coppery brown wing covers. There are six tufts of white hairs on each side of the abdomen. The egg is white, spheroid, and about 1.5 mm in diameter. The last-stage larva is 10.5 mm long and is a typical C-shaped white grub. The distinguishing larval characteristic is on the ventral side of the last abdominal segment, which has two rows of six or seven shorter straight spines arranged in the form of a V.

The Japanese beetle has one generation per year in North America. The larvae overwinter beneath turf from 5 to 15 cm below the soil surface. When the soil warms above 10°C in the spring, the larvae move to within the upper 5 cm of soil, feed for a time, form a soil cell, and pupate. In the South, the beetles begin to emerge from the soil as early as late May. They reach maximum abundance in June, and a few remain after mid-July in central North Carolina. In the North, they begin emerging the last week of June and are often present in great abundance through early September. Females attract males with a pheromone and mate very soon after emerging. The adults mate either in the turf or on plants. The females burrow into the soil, lay eggs, and then exit the soil. Females frequently repeat the feeding–egg laying cycle throughout their life.

Damage

The larvae cause damage to turf by feeding principally on grass roots. Adults are gregarious skeletonizers of foliage. They feed on flowers and fruits of more than 275 plant species. Adults fly, land, and feed actively on warm, sunny days. They prefer to feed on ripe bramble fruit exposed to sunlight. Although they prefer red over black raspberries, they will feed on either. *Rubus* cultivars that bear fruit during peak adult feeding from about late July through mid-September may be severely damaged by adults. Often several beetles feed on a single ripe berry. A secondary injury to bramble crops by Japanese beetles is due to feeding on leaves. In severe cases, foliage exposed to sunlight may be completely skeletonized.

Control

The broad-spectrum insecticide carbaryl is very effective in controlling Japanese beetles and the other scarab beetles mentioned in this section. Care must be exercised, since some formulations are quite toxic to bees.

Both sexes of adults are attracted to Japonilure, an artificial lure mixture consisting of a blend of floral volatiles and a sex pheromone. If traps baited with these attractants are used in high numbers over a wide area, they can reduce populations.

Japanese beetles have several natural enemies. A bacterial disease of the larvae, milky spore disease (caused by *Bacillus popilliae* Dutky), has caused reductions in the number of emerging adults in older infestations. Two imported parasitic wasps, *Tiphia vernalis* Rohwer and *T. popilliavora* Rohwer, consume large numbers of grubs, as do moles, skunks, and birds.

Green June Beetle

The green June beetle, *Cotinis nitida* (L.), is widely distributed in the southeastern United States. Its range extends from the East Coast westward to the 100th meridian in Texas, Oklahoma, and Kansas, and northward beyond the 40th parallel.

The green June beetle larva is often associated with pasture, garden, or lawn areas, and the adults feed in *Rubus* plantings adjacent to where animal manure or compost has been deposited or used to enrich the soil.

Life History and Description

The green June beetle requires 1 year to complete its life cycle. The adult is a robust, colorful metallic green and brown insect, about 25 mm long (Plate 92). Larvae are white scarab grubs that characteristically crawl on their backs. The last abdominal segment of the larva has a raster pattern consisting of two irregular parallel rows of about 15 short setae each. Third-stage larvae overwinter in soil at a depth of 20–30 cm. As the soil warms to above 10°C in spring, the nocturnal larvae may be seen on the soil surface, where they feed on decomposed manure. Hills of oblong pellets (fecal droppings), which are about 1–2 cm high and 5 cm long, are usually found around the vertical burrow of the larva. Starting in mid- to late May, the larvae form oval soil cells from just below the root zone to a depth of 30 cm and then pupate. The pupa molts to an adult, which remains in the soil cell, possibly until an environmental cue, such as heavy rain, triggers its emergence (in Arkansas, from late June to early August). After females emerge, they remain in vegetation near the soil surface or fly to trees. Females attract males via a pheromone and mate very soon after emerging. Afterwards, the females select a suitable turf area, burrow into the soil, and lay eggs. Typically within 1 or 2 weeks after mating, both sexes are attracted by volatiles from damaged ripe fruit. More adults are elicited to form feeding aggregations as a result of the release of an aggregation semiochemical by beetles feeding on fruit. Adult feeding on ripe fruit and oviposition in the soil continues from July to early September.

Damage

The larvae have been reported to cause damage by uprooting and desiccating young turf and grass in pastures, but the major economic damage is caused by the gregarious feeding of the adult beetle on ripe fruit. Adult feeding can cause extensive damage to *Rubus* fruits, and the beetles' foul excrement drips onto adjacent fruit, making it tainted and worthless. In addition, fruit pickers may be startled and distressed by the sudden buzzing flight of adult beetles disturbed from feeding.

Control

The major factors contributing to the natural mortality of green June beetle larvae include soil flooding in the spring (which increases the chances for parasitization by the green muscadine fungus, *Metarrhizium anisopliae* (Metsch.) Sorok., or by nematodes) and dry, hot weather during egg laying and hatching. Other mortality factors include predation by moles and birds and parasitization by the orange-banded digger wasp, *Scolia dubia* Say, and sarcophagid flies.

Control is usually not warranted. However, where the need arises, repeated insecticide applications are required to control the adults.

Rose Chafer

The rose chafer, *Macrodactylus subspinosus* (F.), also known as rose bug, is distributed across much of North America east of the Rocky Mountains. This insect is especially abundant in areas of light sandy soil, where beetles may suddenly appear in large numbers when host plants begin to bloom. The rose chafer feeds on various fruits and ornamental plants, causing considerable damage to flowers, fruit, and foliage. In severe cases, defoliation in *Rubus* plantings can result in reduced yields.

Life History and Description

The adult rose chafer is about 13 mm long, with long, spiny, reddish brown legs, which gradually become darker near the tip (Plate 93). The ungainly beetles have a straw-colored body, reddish brown head, and black undersurface.

Eggs of the rose chafer are oval, white, shiny, and about 1 mm long. Larvae are C-shaped white grubs about 19 mm long. When fully developed, larvae have three distinct pairs of legs, a brown head capsule, and a dark rectal sac visible through the integument. They are found feeding on grass roots in sandy soil and can be identified by a distinctive rastral pattern of two nearly parallel rows of eight or nine short, pointed spines.

Adult rose chafers become active in late May to early June. They often appear suddenly and in great numbers. Beetles feed and mate soon after emerging from the soil. Females lay eggs singly at a depth of 10–15 cm. Beetle activity lasts from 4 to 6 weeks, and the average life span is 3 weeks.

Damage

Despite its common name, the rose chafer attacks the flowers, buds, foliage, and fruit of numerous plants. Unlike the preceding scarabs, rose chafers feed primarily on the white flowers and foliage of *Rubus* spp. Flower buds are often completely destroyed, resulting in little or no fruit production. Feeding activity on various plants may continue for 2–4 weeks. Damage can be especially heavy in sandy areas, which are the preferred habitat for egg laying.

Control

Beetles emerge in mass; therefore, the timing of control methods is important. Foliar applications of insecticides may need to be repeated at weekly intervals or after periods of rain to protect flowers, fruit, and foliage. Because the pupal stage is extremely sensitive to disturbance, plowing or cultivating may be effective in destroying them. Beetles may also fly some distance to attractive plants; therefore, controlling larvae in the immediate vicinity may not preclude damage by adults in that area.

Selected References

Davis, J. J., and Luginbill, P. 1921. The green June beetle or fig eater. N.C. Agric. Exp. Stn. Bull. 242:35.

Domek, J. M., and Johnson, D. T. 1987. Evidence of a sex pheromone in the green June beetle, *Cotinis nitida* (L.) (Coleoptera: Scarabaeidae). J. Entomol. Sci. 22:264-267.

Domek, J. M., and Johnson, D. T. 1990. Inhibition of aggregation behavior in the green June beetle (Coleoptera: Scarabaeidae) by antibiotic treatment of food substrate. Environ. Entomol. 19:995-1000.

Johnson, D. T., and Schaefers, G. A. 1989. Insect and mite pests of brambles. Pages 73-85 in: Small Fruit Pest Management and Culture. D. Horton, R. Bertrand, and G. Krewer, eds. Cooperative Extension Service, College of Agriculture, University of Georgia, Athens.

Ladd, T. L., Klein, M. G., and Tumlinson, J. H. 1981. Phenyl propionate + eugenol + geraniol (3:7:3) and Japonilure: A highly effective joint lure for Japanese beetles. J. Econ. Entomol. 74:665-667.

McLeod, M. J., and Williams, R. N. 1989. Biology and behavior of the rose chafer, *Macrodactylus subspinosus* (F.). Ohio Insect Information. Home, Yard and Garden Facts. HYG-2146-89. OCES/Ohio State University, Columbus.

Miner, F. D. 1951. Green June beetle damage to fall pasture. J. Kansas Entomol. Soc. 24:122-123.

Williams, R. N. 1988. Insect identification and control. Pages 27-38 in: Brambles: Production, Management, and Marketing. R. L. Overmyer and R. C. Funt, coordinators. Bull. 783. Ohio State University, Columbus.

(Prepared by D. T. Johnson and R. N. Williams)

Lygus Bugs

Lygus bugs are inconspicuous plant bugs with sucking mouth parts. They feed on the flowers of a wide selection of plants. They occur throughout North America, Europe, and Asia. Several species may be involved in feeding on raspberry, but the tarnished plant bug, *Lygus lineolaris* (Palisot de Beauvois), has been implicated most frequently. In other parts of the world, Lygus bugs belonging to other genera are commonly found in association with *Rubus*. For example, in the United Kingdom, the common green capsid, *Lycocoris pabulinus* (L.), is most common on raspberry and blackberry.

Life History and Description

In northern North America, the plant bug overwinters as an adult. It generally seeks out protected areas, such as those under leaf litter, mulch, or deep grass. Overwintered adults begin depositing eggs in weed hosts in late April or early May. Depending on the temperature, hatching begins about 2 weeks later. Developing nymphs and young adults generally remain on the low ground vegetation until fruit begins to form in June. At that time the adults move up to the young fruit to feed. There are thought to be two or three generations a year.

Adult plant bugs fly very quickly when disturbed and thus are not readily observed. They are about 6 mm long, oval, somewhat flattened, and greenish brown, with reddish brown markings on the wings. A distinguishing characteristic is a small, yellow-tipped triangle on the back behind the head (Plate 94). The nymphs are pale green when they first hatch and are very small. They may resemble aphids but are readily distinguished by the rapidity of their movement and the lack of posterior tubes (cornicles). As they mature, they begin to take on a brown color and develop a pattern of five dots on their backs.

Damage

Lygus bug injury to raspberry has not been extensively studied. In the United Kingdom, feeding damage to the foliage has been described and is considered unimportant. Feeding on the growing tip of the canes, however, results in branching or death of the canes. In North America, feeding during early fruit development results in failed drupelets and malformed fruit (Plate 95). This injury can cause significant yield reductions in blackberries as well as raspberries. Feeding may occur on the mature fruit, resulting in a whitening of the damaged drupelet.

Control

Information on the control of tarnished plant bugs on *Rubus* spp. is scarce. The value of chemical treatments has been demonstrated for blackberries, and a recent report showed that malathion applied to raspberries immediately after bloom reduced fruit deformities caused by the tarnished plant bug by 5–60%. This translated into a 5–7% increase in berry weight. The need for chemical treatment is determined by knocking the immature and adult insects from blossom and fruit clusters into a flat dish. This should be done after the sun has had a chance to warm up the foliage in mid-morning. Thresholds have not yet been determined, but it is estimated that one or more plant bugs per cluster would warrant insecticide application, particularly in the case of blackberries.

Since egg deposition occurs in many weed species, it is likely that maintenance of weed-free plantings and border areas would aid in reducing plant bug numbers. *Lygus* spp. are parasitized in the egg, larval, and adult stages by a complex of natural enemies. Their effectiveness in significantly reducing pest levels has not been verified.

Selected References

Clancy, D. W., and Pierce, H. D. 1966. Natural enemies of some Lygus bugs. J. Econ. Entomol. 59:853-858.
Hill, A. R. 1952. Observations on *Lygus pabulinus* (L.), a pest of raspberries in Scotland. Pages 181-182 in: Annu. Rep. East Malling Res. Stn. Kent 1951.
Schaefers, G. A. 1980. Yield effects of tarnished plant bug feeding on June-bearing strawberry varieties in New York State. J. Econ. Entomol. 73:721-725.

(Prepared by G. A. Schaefers)

Raspberry Bud Moth

The raspberry bud moth, *Lampronia rubiella* (Bjerkander), or *raspberry moth*, as it is known in the United Kingdom, is found throughout the United Kingdom, northern and central Europe, and eastern Canada, damaging red raspberry and occasionally loganberry.

Life History, Description, and Damage

Adult raspberry bud moths are small, with a wingspan of 9–12 mm, and have attractively marked forewings (Plate 96). The background of the forewings is dark brown and interspersed with many creamy yellow spots. When the insect is at rest, two of these spots are relatively large. The hind wings are uniformly purplish grey. Both sets of wings are fringed with long cilia. The head is a pale fawn color, with well-developed eyes and filamentous antennae. Adult moths fly during sunny days and also at night. The flight period usually coincides with the flowering of raspberry. After mating, the female lays small, ovoid, translucent eggs (0.35 mm long) in the flowers at the junction of the calyx and receptacle. When the larvae hatch after about 10 days, they are translucent, except for a dark, almost black, head capsule. By the late second instar they begin to develop red pigmentation in the body (Plate 97). They initially feed on developing drupelets before tunneling into the receptacle, where they cause little damage. When about 4 mm long, the red larvae migrate to the soil, molt, and hibernate over the winter. In the spring (late April in Scotland) they emerge from hibernation and climb the fruiting canes until they locate an expanding lateral bud, where they feed. They penetrate the bud and tunnel until they reach the pith of the cane. Since the tunnels rapidly become blocked with frass, which conceals the larvae, the only way to confirm their presence is to dissect the bud. Larvae undergo another molt within the bud, and part of the population pupates in the buds. Some of the larva emerge from the buds before pupation and attack other buds or small lateral branches. Those exhibiting this behavior usually have an additional larval stage, and it is these larvae which can cause the most severe damage by destroying many fruit-bearing laterals. Mature larvae spin a silken cocoon in which to pupate. Pupation can occur in the hollowed-out buds or on leaves, canes, or supporting posts and wires. The pupae are initially deep red but rapidly turn brown. Adults emerge in about 3 weeks.

Control

Prior to the introduction of the organochlorine insecticide DDT in the 1950s to control raspberry fruitworm, *Byturus tomentosus* (De Geer), this moth was regarded as a major pest of raspberry in Scotland. Prior to the introduction of DDT, control was achieved by the use of DNOC–petroleum oil or tar oil applied to the base of dormant canes in February or March. With the widespread use of DDT, the moth became rare in commercial plantings and has remained so, even after the withdrawal of DDT. It appears that organophosphorus insecticides that replaced DDT continue to control the moth. However, in the absence of insecticides applied to control other pests, the population can quickly increase. Sanitation practices, such as removing and burning cane debris at the end of the season, should help in reducing raspberry bud moth populations.

Selected References

Hill, A. R. 1952. The bionomics of *Lampronia rubiella* (Bjerkander), the raspberry moth, in Scotland. J. Hortic. Sci. 27:1-13.
Maxwell, C. W. B., and Lord, F. T. 1939. Notes on *Lampronia rubiella*, Bjerk. a raspberry pest new to North America. Rep. Entomol. Soc. Ont. 70:49-51.
Morris, R. F. 1972. Note on the raspberry bud moth, *Lampronia rubiella* (Lepidoptera: Incurvariidae), new to Newfoundland. Can. Entomol. 104:917.

(Prepared by S. C. Gordon)

Picnic Beetles

The picnic beetles, *Glischrochilus quadrisignatus* (Say) and *G. fasciatus* (Olivier), belong to a family of beetles known as sap beetles, or Nitidulidae. Although the name *picnic beetle* is not an officially accepted common name, it is commonly used to describe these two insects. The two species are treated together here because they are similar in size, appearance, and the injury caused to *Rubus* spp. The picnic beetles are widely distributed throughout northeastern North America. *G. quadrisignatus* was found in Europe after World War II and is now in several countries neighboring Germany and Czechoslovakia. The name *picnic beetle* seems appropriate, since they are nuisances at picnics—attracted to beer, potato salad, bread, fruit salad, and many other foods.

Life History and Description

Picnic beetles overwinter in the adult stage in leaf litter, under bark, in downed corn stubble, or in almost any organic matter that is in some stage of decomposition and could provide shelter during the winter months. When temperatures reach 15–20°C for several days in a row, adult activity is initiated. However, the beetles return to protected places if ensuing temperatures fall to 10°C or lower. Upon emergence from their overwintering sites, they seek a source of food, such as mushrooms, fungal mats, downed corn, sap from plant wounds, or pollen, and begin feeding. Eggs are laid on many forms of decomposing organic materials. In corn production areas, downed corn ears are an important substrate for picnic beetle aggregation, egg laying, and larval development. Beetle populations build to a peak by midsummer. There is only one generation produced per year.

The fully-grown larvae are 3.4–9.2 mm long and white, with a brown head capsule ranging in width from 0.9 to 1 mm. The adult beetles are from 4 to 7 mm long and shiny black, with two yellow-orange spots on each wing cover (Plate 98). Wing covers extend over the entire abdomen. The antennae of adult picnic beetles are knobbed on the tips.

Damage

Adult sap beetles bore into ripe and overripe raspberries at harvest time. They eat a portion of the fruit and may lay eggs. Their damage generally leaves the fruit undesirable and may be extensive enough to prohibit its sale in some areas. In addition, they can be a major contaminant in harvested fruit. Picnic beetles also disseminate fungi and bacteria that cause fruit rots. Late-maturing raspberry cultivars tend to be more prone to beetle infestations than early-maturing cultivars. This is probably due to a larger beetle population in late summer.

Control

Chemical control of picnic beetles is generally not acceptable. Insecticide applications near harvest leave visible residues, which are objectionable. In addition, the beetles are usually well protected from insecticide application, because they are wedged between the receptacle and the berry.

Sanitation is an important control practice. Picnic beetle adults are attracted to ripe, and particularly overripe and fermenting, fruit. Damaged, overripe, and fermenting fruit should not be allowed to accumulate in the field, since it serves as a powerful attractant for the beetles. To prevent contamination after harvest, picked berries should be removed from the field as soon as possible or covered. Damaged berries should be removed from the planting and destroyed.

Selected References

Connell, W. A. 1956. The Nitidulidae of Delaware. Univ. Del. Agric. Exp. Stn. Bull. 318. 67 pp.
Foott, W. H., and Timmin, P. R. 1971. Importance of field corn as a reproductive site of *Glischrochilus quadrisignatus* (Say) (Coleoptera: Nitidulidae). Proc. Entomol. Soc. Ont. 101:73-75.
Luckmann, W. H. 1963. Biology and control of the picnic beetle, *Glischrochilus quadrisignatus* (Say). Proc. N. Cent. Br. Entomol. Soc. Amer. 18:38-39.
Miller, K. V., and Williams, R. N. 1981. An annotated bibliography of the genus *Glischrochilus* Reitter. Ohio Agric. Res. Dev. Cent. Res. Bull. 266. 65 pp.

(Prepared by R. N. Williams)

Raspberry Fruitworms

The larvae of the raspberry fruitworm, *Byturus unicolor* Say, in North America, and the raspberry beetle, *B. tomentosus* (De Geer), in Europe, usually cause more damage on raspberries than the adults. However, the adult beetles are also capable of causing injury to unopened buds, unfolding leaves, and open flowers. In North America, the raspberry fruitworm prefers red and purple raspberries, but in Europe they also infest blackberries (both wild and cultivated) and hybrid berries, such as tayberry and loganberry.

Life History and Description

Adult beetles emerge from the soil in late April and early May. In North America, they begin feeding along the midrib of partially folded leaves and are usually found on the midrib of young leaves. In Europe, this type of damage is usually confined to primocane leaves. Later in the season, the adults seek protection between the flower buds. They attack these buds and make large entrance holes to feed on the floral parts. In some parts of Europe, adult beetles may migrate to adjacent rosaceous flowers (such as apple, pear, or hawthorn) to feed before returning to red raspberry when the flowers open.

The females deposit their eggs most commonly on swollen, unopened flower buds. However, at times they may be laid inside buds or on developing fruit. The grayish white eggs (approximately 1 mm long) hatch after a few days, and the larvae commonly bore through the bud and enter the receptacle torus, where they begin to tunnel. As the larvae increase in size, the tunnel is made larger and ultimately becomes a groove in the receptacle adjacent to the berry. (In the case of *B. tomentosus*, the cycle is slightly different. Eggs are laid attached to floral parts; the young larvae initially feed on basal drupelets until the torus softens, and then they begin to tunnel.) When infested fruit is picked, the larvae often remain attached to the cuplike interior of the fruit and thus become a contaminant in harvested berries. The larvae remaining on the receptacle soon drop to the ground, where they pupate and remain over winter. The fully grown larvae are slender, 5.75–6 mm long, and 0.53 mm wide (Plate 99). They are nearly cylindrical, tapering towards either end. Each body segment has sparse, light-colored, stiff hairs arranged in two transverse rows along the length of the body.

The adult beetle ranges in length from 3.7 to 4.5 mm. Beetles are oblong to oval, convex above, and dull yellow to pale brown to reddish brown (Plate 100). When viewed under a hand lens, one can see that they are covered with tiny hairs. Antennae are 11-segmented and terminate in a three-segmented club.

Damage

Generally the first indication of a problem is the presence of small, yellowish white larvae adhering to harvested fruit. However, numerous signs occur earlier in the season. Early-season infestations are suspected if longitudinal holes in the foliage leave a tattered appearance. Such foliar injury is caused by adults feeding on unfolding leaves. If sufficient feeding occurs, leaves may become skeletonized. As flower buds appear, they are also attacked by the adult beetles, which feed on the insides; numerous beetles may destroy the entire flower cluster. Fruitworm larvae attack raspberry receptacles and, at times, the carpels of the berry. In tunneling through the receptacles, the larvae cause extensive damage, often loosening berries so that they fall off prior to harvest. Contamination of harvested fruit by fruitworm larvae can be a serious problem.

Control

When fruitworms are a problem in commercial plantings, they are usually controlled by applications of a suitable organophosphorus insecticide. Prebloom sprays are applied as flower buds appear and again before flowers open. In the United Kingdom, a single application at the late green fruit stage often gives adequate control. Since fruitworms fall to the soil in late July, fall-fruiting cultivars generally do not have problems caused by this insect in North America. However, in some years in the south of England, fall-fruiting cultivars may be heavily infested.

Selected References

Baker, W. W., Crumb, S. E., Landis, B. J., and Wilcox, J. 1947. Biology and control of the western raspberry fruitworm in western

Washington. Wash. State Coll. Agric. Exp. Stn. Bull. 497 pp.

Goodwin, W. H. 1909. The raspberry *Byturus*. Ohio Agric. Exp. Stn. Bull. 202:173-186.

Schaefers, G. A., Labanowska, B. H., and Brodel, C. F. 1978. Field evaluation of eastern raspberry fruitworm damage to varieties of red raspberry. J. Econ. Entomol. 71:566-569.

Taylor, C. E. 1971. The raspberry beetle (*Byturus tomentosus*) and its control with alternative chemicals to DDT. Hortic. Sci. 11:107-112.

Taylor, C. E., and Gordon, S. C. 1975. Further observations on the biology and control of the raspberry beetle (*Byturus tomentosus* (Deg.)) in eastern Scotland. J. Hortic. Sci. 50:105-112.

(Prepared by R. N. Williams and S. C. Gordon)

Yellowjackets

Yellowjackets are large yellow and black or white and black wasps in the family Vespidae (Plate 101). All are carnivorous and catch insects with which to feed their larvae, and a few species have a propensity to scavenge for protein and carbohydrates. All species have annual colonies produced by a single queen, who builds a nest each year in the spring. North America has the largest number of species with 19, but yellowjackets are nearly ubiquitous throughout North America, Europe, New Zealand, and other *Rubus*-growing areas. The term *yellowjacket* seems to have originated in the New World; Europeans simply call them wasps. Most of the troublesome, scavenging species belong to the genus *Paravespula* (syn. *Vespula vulgaris*). True hornets are closely related, but they are larger old-world wasps in the genus *Vespa*.

Yellowjackets are serious nuisance pests in red raspberry plantations. The adults are attracted to ripe or injured fruit as a source of moisture and sugar and while feeding are an extreme annoyance and danger to pickers. Prompt harvesting of ripe berries and "clean" picking practices will help decrease the fruit's attractiveness to yellowjackets.

In the Pacific Northwest, the most troublesome species, *P. pensylvanica* (Saussure), can be managed by using traps with the synthetic lure heptyl butyrate. Unfortunately, this is not an effective lure for *P. germanica* (F.) and other species of the *Paravespula* group that are distributed worldwide.

An alternative control for any species of *Paravespula* uses a fish with the sides cut to expose flesh. The fish is suspended above a container of water containing a wetting agent. The yellowjacket workers attempt to fly away with a large piece of fish, fall into the water, and drown.

Selected References

Akre, R. D., and Antonelli, A. L. 1989 (rev.). Yellowjackets and paper wasps. Wash. State Univ. Ext. Bull. 0643. 6 pp.

Akre, R. D., Greene, A., MacDonald, J. F., Landolt, P. J., and Davis, H. G. 1980. Yellowjackets of America north of Mexico. U.S. Dep. Agric. Agric. Handb. 552. 102 pp.

Akre, R. D., and MacDonald, J. F. 1986. Biology, economic importance and control of yellowjackets. Pages 353-412 in: Economic Impact and Control of Social Insects. S. B. Vinson, ed. Praeger, New York.

MacDonald, J. F., Akre, R. D., and Keyel, R. E. 1980. The German yellowjacket (*Vespula germanica*) problem in the United States (Hymenoptera: Vespidae). Bull. Entomol. Soc. Am. 26:436-442.

(Prepared by R. D. Akre and R. N. Williams)

Strawberry Bud Weevil (Clipper)

The strawberry bud weevil (*Anthonomus signatus* Say), or "clipper," is a native of eastern North America that attacks strawberry as well as *Rubus* spp. and blueberry. These insects clip off buds in early spring, at times causing considerable loss in fruit production. The adult is small (3 mm long) and dark reddish-brown, with a head that is prolonged into a slender, curved snout about half as long as the body. They overwinter under organic matter in protected places along fence rows and in woodlots. As soon as weather permits they move to the fruiting buds, where they first feed on immature pollen by puncturing the blossom buds with their long snouts. The female deposits a single egg inside the nearly mature bud and then girdles it, so that the bud falls to the ground or hangs on by a thread (Plate 102). The larva develops inside the bud, reaching maturity in 4-5 weeks. Only one brood is produced each year. Small round holes in the flower petals and severed or dangling buds are indications of clippers in a planting. Insecticide applications when symptoms are first observed are used for control.

(Prepared by R. N. Williams)

Insects and Mites That Damage Foliage

Spider Mites

Several species of tetranychid mites have been reported to damage *Rubus* spp. worldwide. Probably the species most commonly referred to is the two-spotted spider mite, *Tetranychus urticae* Koch. The other species that attack *Rubus* have similar life habits and cause similar symptoms. These spider mites are general feeders able to survive and reproduce on hundreds of cultivated and wild plant species. The spider mite problem is rare on *Rubus* grown in cool areas, such as Scotland, but it is common in warmer and drier climates.

Life History and Description

Spider mites are arthropods that belong to the family Tetranychidae. They resemble tiny spiders and have eight legs when mature (Plate 103). They are approximately 0.5 mm long when fully grown. They vary in color from straw to green or red, with two darker spots on the upper surface of their body. Fertile female spider mites or young nymphs overwinter in the folds of old leaves on the ground or in cracks and crevices of the canes and supporting stakes. Mites colonize *Rubus* in the early spring as foliage begins to form. Eggs are deposited on the undersurfaces of *Rubus* leaves. As the season advances, progressive generations develop and colonize new foliage as it appears.

Damage

Mite injury is generally greatest during dry, hot seasons and to plants under stress. The first symptom on the upper leaf surface is a light, stripped appearance where the chlorophyll has been removed. The feeding process removes sap from leaves, causing them to gradually turn yellow, silver, or bronze (Plate 104). Webbing is usually present on the lower surface of the leaves. Heavy infestations stunt canes and leaves. Severe infestations may bronze the leaves or cause early leaf drop. Severe damage to fruiting canes may reduce yield and fruit quality. Primocane feeding stunts cane growth, thus reducing crop potential for the following season. Economic damage also occurs when mite populations cause premature defoliation of plants, reducing winter hardiness. Dusty conditions contribute to large mite populations. The early-season application of foliar pesticides, particularly carbamate and some synthetic pyrethroid insecticides, tends to increase mite populations, probably by killing beneficial predators or stimulating female mites to lay more eggs.

Control

Routine scouting for mites on *Rubus* leaves is recommended and should be conducted weekly. Randomly selected leaves should be inspected for the presence of mites. The first clue is webbing on the undersurface of the leaves. To see the mites move without magnification, a leaf with webbing may be removed and held in sunlight. To count mites, a 10× hand lens should be used, and 100 leaves should be checked from across the *Rubus* planting; if there is an average of 10-15 mites per leaf, most authors recommend miticide treatment. If mites are a perennial problem, a dormant spray before budbreak is often necessary.

Predaceous mites can contribute significantly to the overall population suppression of spider mites in *Rubus* fields. Leaf samples can be taken early in the season to verify the presence of these mites. Adopting certain agricultural practices that foster the development of predaceous mite populations will contribute to the control of spider mites. These include selecting insecticides (when needed for early-season insect control) with the least toxicity to predaceous mites, following an irrigation schedule to assure adequate water for plants, and eliminating or greatly reducing road dust. Broadleaf weed control will remove alternate hosts on which spider mites may develop and increase during the early part of the season.

Red raspberries appear to be among the most susceptible of the *Rubus* spp. Early in the season, miticides are occasionally used to effectively reduce large populations. However, it is doubtful that spider mite infestations in red raspberries require chemical control after mid-September in the more northern latitudes.

Selected References

Doughty, C. C., Crandall, P. C., and Shanks, C. H. 1972. Cold injury to red raspberries and the effect of premature defoliation and mite damage. J. Amer. Soc. Hortic. Sci. 97:670-673.

Jepson, L. R., Keifer, H. H., and Baker. E. W. 1975. Mites Injurious to Economic Plants. University of California Press, Berkeley. 614 pp.

(Prepared by G. C. Fisher)

Dryberry Mite (Raspberry Leaf and Bud Mite)

The dryberry mite, *Phyllocoptes gracilis* (Nalepa), also known as the raspberry leaf and bud mite in the United Kingdom and the raspberry mite in Europe, occurs in many of the raspberry-growing areas of the world. Mites have been reported from the Pacific Northwest, Great Britain, and continental Europe west of the Ural Mountains. They colonize both cultivated and wild red raspberries, blackberries, and hybrid berries, such as tayberry, loganberry, Himalaya berry, and thimbleberry.

Life History and Description

The dryberry mite belongs to a group of microscopic mites known as the eriophyid mites. This mite, like others in the group, has two alternating, morphologically distinct forms. The overwintering female form (males are absent) is called the deutogyne. It is light brown and is induced to lay eggs after exposure to a period of cold weather. The summer form (of both males and females) is called the protogyne. It is translucent and whitish yellow and lays eggs after mating.

In winter, pinkish brown colonies of up to several hundred individuals can be found under the outer bud scales (Plate 105) or, occasionally, in cracks of bark on primocanes. The size of the colony partly depends on the cultivar, since considerable variation in varietal susceptibility has been demonstrated. Although mites can withstand very low temperatures, overwintering mortality is usually high, probably from mites being crushed between the bud scales or from desiccation. In the spring, as buds open, overwintered females emerge to feed on the expanding foliage and lay eggs among the hairs on the undersides of leaves. After hatching, larvae of the first summer generations feed on epidermal cells on the undersides of leaves on fruiting canes. Adults and larvae of these generations are wormlike, with two pairs of legs close to the head and a pair of long, whiplike setae to the rear. Adults measure about 0.15 mm long. As leaves expand, irregular yellow blotches soon develop, and in severe infestations virtually all the leaves on the plant may be affected. In some red raspberry cultivars and in tayberry, leaf size may also be reduced (Plate 106). The terminal growing point of primocanes may be damaged, and the proliferation of lateral buds results in weak, branched canes, which are difficult to tie in during winter. The yellowing of leaves is frequently confused with diseases caused by viruses. This mistaken diagnosis can result in premature destruction of the plantation. In plants with heavily pigmented leaves, such as tayberry, the early symptoms of mite feeding may be difficult to detect.

Damage

The mites mainly infest the leaves on fruiting canes until the fruit is harvested, and then their numbers rapidly decline. With this decline there is a corresponding increase in mite numbers on leaves of primocanes, which reaches a peak in September. When mite populations are high, they will colonize and feed on the developing fruit. The degree of fruit damage depends on the host. In red raspberry, mite damage to developing fruit leads to premature ripening of some of the drupelets, resulting in badly misshapen fruit that is difficult to harvest. Fruit of heavily infested loganberry dries out soon after petal fall and dies before ripening—thus the name *dryberry disease*. Eriophyid mites are thought to migrate during the summer on wind currents or to be carried by flying insects. These are probably the main methods of spread. The exact distance of migration is unknown. Some nursery stocks of red raspberry in Scotland were shown to have a small population of mites overwintering in the buds; thus, movement on nursery stock could provide a means for long-distance spread. However, the chance of spread by nursery stock should be low, because it is normal commercial practice to cut the canes close to the soil surface after planting. Just prior to leaf fall, light brown females are produced, and they move into the buds to overwinter. In mild winters, mites may feed on the green tissue of the outer bud scales, but damage is usually slight and insufficient to prevent bud opening in the spring.

Control

Once a plantation becomes infested with dryberry mites, control can often be difficult without the use of miticides. In the United Kingdom, good control has been achieved with the systemic insecticide vamidothion applied to infested canes in the spring. The standard broad-spectrum miticides have proven to be ineffective, and dryberry mite populations are often greater after treatment with such compounds. Sulfur-containing compounds have been used with varying degrees of success. The relative inefficiency of the standard broad-spectrum miticides may be due to their effect on predators, especially other mite species. In the United Kingdom, the predatory mite *Typhlodromus pyri* Scheuten is commonly found in dryberry mite–infested plantations and is frequently in hedgerows, especially on European blackberry, *Rubus fruticosus* L. Similarly, *Phytoseius macrophilis* (Banks) was reported to reduce dryberry mite numbers in Yugoslavia.

Since dryberry mite numbers are greatest on plants growing in sheltered locations, the risk of damage could be lessened by avoiding such sites.

Selected References

Breakey, E. P. 1945. *Phyllocoptes gracilis*, a pest of red raspberry in the Puyallup Valley. J. Econ. Entomol. 38:121-122.

Dobrivojevic, K., and Petanovic, R. 1985. The eriophyid raspberry leaf miner *Phyllocoptes gracilis* (Nal.) (Eriophyoidea, Acarina), a little known pest in Yugoslavia. (In Serbo-Croatian.) Zast. Bilja 36:247-254

Domes, R. 1957. Zur Biologie der Gallmilbe *Eriophyes gracilis* Nalepa. Z. Angew. Entomol. 41:411-424.

Gordon, S. C., and Taylor, C. E. 1976. Some aspects of the biology of the raspberry leaf and bud mite (*Phyllocoptes* (Eriophyes) *gracilis* Nal.) Eriophyidae in Scotland. J. Hortic. Sci. 51:501-508.

Gordon, S. C., and Taylor, S. C. 1977. Chemical control of the raspberry leaf and bud mite, *Phyllocoptes gracilis* (Nal.) (Eriophyidae). J. Hortic. Sci. 52:517-523.

(Prepared by S. C. Gordon)

Redberry Mite

The redberry mite, *Acalitus essigi* (Hassan), is a perennial pest of both wild and cultivated blackberries in western North America and in Europe. This eriophyid mite is microscopic; in order to see the mites well, a 20× magnifier is useful.

Life History, Description, and Damage

Redberry mites overwinter on blackberry canes among bud scales or deep in buds. During spring and summer the mites remain near the axillary buds of the plants. As the blossoms develop and fruit forms, the mite populations increase among the berry drupelets, particularly near the bases, and around the core of the fruit.

Significant damage to commercially grown blackberries, particularly evergreen blackberry, can result. Fruit infested with redberry mites does not develop normally colored drupelets. Severely infested drupelets usually remain hard and green or bright red. This is often referred to as redberry, the intensity of which is evidently related to the numbers of mites present. Later-maturing blackberry cultivars generally display the greatest amount of damage.

Control

If not controlled, this mite can spread from a few canes in one season to a sizeable portion of plantings the following season. Once high populations of redberry mites become established in commercial blackberry plantings, the application of registered sulfur compounds to the plants remains the most effective control. Both contact and vapor action of the material affect mortality. Combinations of fall, winter, and spring applications of lime sulfur, lime sulfur and oil, and wettable sulfur generally provide acceptable control. Large populations are best controlled with at least two applications—the first in the fall or late winter, followed by a second in the spring. In the United Kingdom, control is obtained by up to three applications of the miticide-insecticide endosulfan before flowering. Annual applications are rarely required.

Because of the extremely small size of the mite and the fact that it is well secluded, the use of power sprayers capable of developing high pressure at the pump is recommended. Low-pressure low-volume sprayers generally do not provide satisfactory control of redberry mite.

Selected References

Breakey, E. P., and Brannon, D. H. 1946. The control of the blackberry mite. Wash. State Univ. Ext. Serv. Ext. Bull. 346. 4 pp.

Edwards, W. O., Gray, K. W., Wilcox, T., and Mote, D. C. 1935. The blackberry mite in Oregon. Oreg. Agric. Exp. Stn. Bull. 337. 33 pp.

Jepson, L. R., Keifer, H. H., and Baker, E. W. 1975. Mites Injurious to Economic Plants. University of California Press, Berkeley. 614 pp.

(Prepared by G. C. Fisher)

Raspberry Aphids

Aphids, insects with sucking mouthparts, are capable of transmitting plant viruses. The principal vector of the raspberry mosaic virus complex in North America is the larger raspberry aphid, *Amphorophora agathonica* Hottes (Plate 107). The species responsible for transmission of the raspberry leaf curl virus is the smaller aphid, *Aphis rubicola* Oestlund (Plate 108). These aphids occur naturally throughout most of the temperate regions of North America. Because of their ability to transmit virus diseases, they represent a major constraint to raspberry production. *Amphorophora sensoriata* Mason is common in the eastern United States. This very pale, slightly bluish green aphid colonizes canes of black raspberry. In Europe, several species are found on raspberry. The most important species is the large raspberry aphid, *Amphorophora idaei* (Börner), a vector of at least four viruses. The small raspberry aphid, *Aphis idaei* van der Goot, tends to be more common in warmer areas of Europe. This species is a vector of raspberry vein chlorosis virus. Other European species have not been implicated as virus vectors. The blackberry aphid (*Amphorophora rubi* Kaltenbach) is also common in Europe and is morphologically indistinguishable from *A. idaei*. It does, however, have a different chromosome number.

Life History and Description

Aphids overwinter in the egg stage. Eggs are deposited in late fall in bud axils on the primocanes or on the underside of old leaves and hatch in the spring when leaves first emerge. A few winged forms occur during the first generation but are scarce in succeeding generations. It is the winged forms that are primarily responsible for the transmission of viruses from crop to crop, but transmission between neighboring plants may occur by walking aphids. All individuals in the summer generations are parthenogenic, do not mate, and give birth to living young instead of laying eggs. As many as 20 generations may occur during the season. As the length of days shortens in the fall, winged males and sexual females are produced, and mating occurs. After mating, the female deposits overwintering eggs on the canes.

In North America and Europe the larger raspberry aphids are the most important pest species. Although both species are similar, the larger North American raspberry aphid is yellow-green, while the European large raspberry aphid tends to be shiny and light green. Four biotypes of the European large raspberry aphid have been identified, but only two are important at the present time. Both of these biotypes are soft-bodied, usually wingless insects, frequently seen feeding on the underside of new raspberry foliage or—as in the case of the large raspberry aphid—on the apex of the primocane. Compared to most aphid species they are quite large, sometimes exceeding 3–4 mm in length.

Aphids are further characterized by a pair of tubes (cornicles) extending posteriorly from the abdomen and by tubular sucking mouthparts.

Damage

Except under conditions of extremely high populations, little damage results from direct aphid feeding. The majority of damage caused by aphids results from their ability to vector or transmit the above-mentioned viruses. Mosaic is the most important of the virus diseases attacking brambles, with yield reductions in excess of 50% attributed to it.

Control

Raspberry aphids are attacked by several parasites and predators. However, under conditions of cool temperatures, the absence of driving rains, and abundant soil moisture, dense populations may develop. Heavy rain and periods of high temperature adversely affect population levels.

Raspberry aphids prefer to feed on the more succulent tissues near the cane tips. Scouting is most effectively conducted by examining the tips of canes early in the season, generally during May and June. In North America, if counts exceed two aphids per cane tip, treatment with an appropriate insecticide may be required to help reduce the within-field spread of viruses. In Scotland, specific insecticidal control is usually unnecessary, but in the warmer areas of the United Kingdom and in some parts of Europe, systemic organophosphorus insecticides, or "winter washes," may be required to prevent damage.

The best method of controlling aphid transmission of *Rubus* viruses in North America is to plant aphid-resistant cultivars, such as Canby, Lloyd George, Royalty, and others; in Europe, it is best to plant cultivars that contain the resistance genes A_1 and A_{10}, such as Delight, Glen Moy, Glen Prosen, Joy, and Autumn Bliss. Resistance to aphids is increasingly emphasized in *Rubus* breeding programs, and culturally acceptable cultivars are becoming available. The resistance-breaking strain of the European large raspberry aphid (biotype 2) has become common in the United Kingdom. This strain is capable of colonizing raspberry genotypes containing gene A_2.

Prevention of high aphid populations may also be accomplished with the application of granular systemic insecticides at planting and the use of foliar sprays when necessary. Granular systemic insecticides are not permitted in the United Kingdom.

Selected References

Birch, A. N. E., and Jones, A. T. 1988. Levels and components of resistance to *Amphorophora idaei* in raspberry cultivars containing different resistance genes. Ann. Appl. Biol. 113:567-578.

Blackman, R. L., Eastop, V. F., and Hills, M. 1977. Morphological and cytological separation of *Amphorophora* Bucktan feeding on European raspberry and blackberry (*Rubus* spp.). Bull. Entomol. Res. 67:285-299.

Schaefers, G. A. 1988. Studying aphid-resistant brambles. Am. Fruit Grow. 108:8.

(Prepared by G. A. Schaefers and S. C. Gordon)

Leaf Rollers

Two species of leaf-rolling caterpillars, the orange tortrix (Plate 109), *Argyrotaenia citrana* (Fernald), and the oblique-banded leaf roller, *Choristoneura rosaceana* (Harris), commonly infest *Rubus* spp. Both species can be important contaminants in mechanically harvested *Rubus* fruits. Of the two species, the orange tortrix is the primary leaf roller contaminant in red raspberries at harvest. Occasionally oblique-banded leaf roller larvae are contaminants on trailing berries. As expected, machine-harvested berries have much greater problems with leaf rollers as contaminants than do those harvested by hand. Several species of leaf rollers, or tortrix moths, are found in Europe, but generally they cause little damage in commercial plantings; however, damage has been reported from New Zealand.

Life History and Description

The orange tortrix has two or three generations per year. The larvae feed and overwinter on many plant species in and surrounding *Rubus* fields, but the bulk of the overwintering population occurs within commercial *Rubus* fields. This leaf roller overwinters in a wide range of larval stages (Plate 110). Larvae are usually found wrapped in dead leaves on canes tied to trellises (Plate 111). During the winter and spring months, the larvae may feed on dead leaves, living canes, and buds. In the spring, summer, and fall, feeding occurs primarily on terminals between webbed leaves. However, the larvae may also feed on ripe fruit. On red raspberry, larvae may be found within the cup of the berry; on boysenberry, they may bore into the fruit through the calyx. Peak emergence of adults in the spring varies from late April to mid-May. It is this flight that produces the larvae that can become contaminants of fruit at harvest.

The oblique-banded leaf roller has two generations per year. It overwinters as an inactive larva on canes under loose bud and bark scales. It has a large host range and occurs on many commercially grown plants. Because populations of its larvae develop later in the year than those of the orange tortrix, the oblique-banded leaf roller is usually more of a contamination problem on trailing berries because of the later harvest of these fruits.

Damage

Both species cause little direct damage to the fruits and canes. Substantial economic loss can occur when larvae become contaminants of the fruit at harvest.

Control

Many beneficial insects and spiders play an important role in reducing leaf roller populations in commercial *Rubus* fields. The mortality of eggs and small larvae stages can be quite high. Parasitization rates as high as 60% can occur in overwintering orange tortrix populations. Even during the summer, parasite levels can be high for both species of leaf rollers. Unfortunately, all life stages of the common parasites and predators are sensitive to insecticides applied to control leaf rollers. Insecticide applications probably diminish the ability of naturally occurring biological control to suppress leaf roller populations to noninjurious levels.

Pheromone traps offer an approach to monitoring leaf rollers. They have been used successfully in commercial *Rubus* plantings in the Pacific Northwest for nearly a decade to decide whether and when leaf roller control is necessary. In Oregon, three classes of catches in pheromone traps for the orange tortrix in red raspberry fields have been established to determine the need for an insecticide application prior to harvest. These classes are based on the peak catches of moths per week as well as the total moth catch from first flight until bloom. A degree-day model also exists to facilitate optimum timing of insecticide sprays to control the larvae.

Certain cultural practices have been shown to reduce orange tortrix infestations. In fields where this leaf roller has been a problem, proper cane training can reduce the overwintering populations of larvae. Timely removal and destruction of old canes in the fall, delaying trellising until most leaves have dropped, and loosely tying canes to the trellis substantially contribute to leaf roller mortality during winter.

Leaf roller larvae can be monitored from late March through April in *Rubus* plantings. Inspections at this time will give an indication of the need to initiate control methods later in the season. Sampling is best accomplished by prying off dead leaves wrapped to the canes and inspecting leaf rolls and webbed leaf terminals for larvae. A critical time to resample the fields is from 10 days to 1 week before harvest, to determine if insect contamination at harvest will be a problem. This is accomplished by placing a white cloth on the ground beneath the canes and shaking them vigorously. Leaf roller larvae will drop down and can readily be seen against the light background. Even if larvae are present only in moderate numbers, an insecticide application is usually recommended for mechanically harvested berries.

The oblique-banded leaf roller is generally not a problem on red raspberry, because of the crop's relatively early harvest in relation to larval development. However, it can be a contaminant on trailing berries. The need to control its larvae is determined by the presence of the larvae, past experience with this pest in a given field, and pheromone trap catches before and during the bloom period.

Selected References

Charles, J. G., White, V., and Cornwell, M. A. 1987. Leafroller (Lepidoptera: Tortricidae) damage to buds of raspberry canes in New Zealand. N.Z. J. Exp. Agric. 15:491-496.

Coop, L., Knight, A., and Fisher, G. 1989. Parasitism of orange tortrix on caneberry, *Rubus* spp., in western Oregon and Washington. J.

Entomol. Soc. B.C. 86:65-67.

Knight, A. L., and Croft, B. A. 1986. Larval survivorship of *Argyrotaenia citrana* (Lepidoptera: Tortricidae) overwintering on small fruits in the Pacific Northwest. J. Econ. Entomol. 79:1524-1529.

Knight, A. L., LaLone, R., Fisher, G. C., and Coop, L. B. 1988. Managing leafrollers on caneberries in Oregon. Oreg. State Univ. Ext. Circ. 1263. 8 pp.

(Prepared by G. C. Fisher)

Climbing Cutworms

Two of the most common species of climbing cutworms in North America are the spotted cutworm, *Xestia* (L.) spp., and the variegated cutworm, *Peridroma saucia* (Hübner). In Scotland, a similar species, the double dart moth, *Graphiphora augur* (F.), has recently become an important but local pest of red raspberry. Climbing cutworms are the larvae of moths in the family Noctuidae.

Life History and Description

The spotted cutworm and the double dart moth overwinter in the larval stage. Insect development is completed in late spring, and the adults lay eggs on the leaves of *Rubus* plants. There are usually two or three generations per year of the spotted cutworm, but only one of the double dart moth. Populations of cutworm larvae may be very sporadic. They can be present at damaging levels in some fields or in an area for several years, and then the population can drop to insignificant levels for many years.

Damage

Damage occurs in two ways. Overwintering larvae climb up the canes at night during budbreak in the spring and eat the swollen buds and new shoots (Plate 112). When populations are high, larvae may completely defoliate portions of a field. The second-generation larvae of the spotted cutworm usually occur near the beginning of harvest in late June or early July. Mechanical harvesters often knock off cutworms, along with ripe raspberries, causing contamination of the harvested berries.

Control

Applications of insecticide are the main means of control. Small, immature larvae are much easier to control than those in later instars, but in Scotland, control of the double dart moth is only practical in early spring, when larvae become active. The best control is achieved by application of insecticides after dark, when the insects are actively feeding. Since cutworms are sporadic in their appearance in red raspberry fields, sprays should not be applied unless a problem exists. Fields should be inspected for damaged buds and shoots at budbreak. Monitoring of a field should be conducted methodically, since only a portion of it may be seriously infested. Cutworms are nocturnal, so it is necessary to search for them at night with the aid of a light. At harvesttime, growers may not be aware of their presence until the mechanical harvester has begun dislodging them along with the berries. Larvae that are picked out of the harvested berries by workers should not be dropped onto the ground, because they will climb back onto the plants to be "harvested" again. They can be killed by dropping them into a pail of water containing detergent.

Selected Reference

Gordon, S. C., McKinlay, R. G., Riley, A. M., and Osborne, P. 1988. Observations on the biology and distribution of the double dart moth (*Graphiphora augur* (Fabricius) in Scotland, and on damage caused by the larvae to red raspberry. Crop Res. 28:157-167.

(Prepared by C. H. Shanks, Jr.)

Blackberry Psyllid

The blackberry psyllid, *Trioza tripunctata* Fitch, has been reported in North America from Ontario to Florida and west to Iowa—wherever *Rubus* and conifers occur together. Psyllids caused considerable damage to commercial blackberry production in the 1920s, particularly in New Jersey, but since then the psyllid has been considered to be a minor pest. Reports of damage have increased with the growing popularity of semierect thornless blackberries in the 1980s.

Life History and Description

Adult blackberry psyllids are approximately 3.5–4 mm long and yellow to brown, with distinctly striped wings (Plate 113). They are easily visible as they feed at shoot tips. They lay narrow, white eggs that are approximately 0.3 mm long. The eggs turn yellow shortly before they hatch. After emerging from the egg, the insect has five nymphal instars. The first four are white to pale yellow, and the oldest nymphs are blue-green and mottled with brown. All nymphs secrete waxes in several different forms. These include wax tubes, flat wax filaments, and the delicate wax "hairs" that cover their bodies. The wax structures break off easily and collect within curled leaves or depressions in leaves to form moldy-appearing masses, which conceal the nymphs.

Blackberry psyllids have one generation a year. Adults spend the winter on evergreen conifers, including pines, spruces, cedars, and hemlocks. No damage has been observed on their winter hosts. Mating occurs from November through March.

Females migrate to blackberries in spring (mid-May in mid-Atlantic states), when new canes are 30–60 cm tall. The edges of blackberry patches closest to overwintering sites are generally most heavily infested, and once a female has landed on a shoot, she remains sedentary and seldom moves. Females feed for about a week before beginning to lay their eggs. They may live as long as a month, laying about 100 eggs. The eggs are usually laid on the underside of young leaves, but occasionally on petioles or small shoots. They hatch in about 10 days. Young nymphs feed and develop throughout the summer and early fall, generally staying in the same feeding sites on the underside of curled leaves. The oldest nymphs often move to stems or shoot tips. Adults emerge in fall (mid- to late October in mid-Atlantic states) and migrate to conifers by early November.

Damage

Blackberry psyllid damage is often mistaken for a plant disease, since growth distortion continues for some time in the absence of any insects. Their feeding causes abnormal plant growth, including leaf curl, shortened internodes, abnormal leaf color, and stunted, misshapen canes (Plate 114). Growth distortion may be seen in the blackberry field within a week of psyllid arrival. Leaf curling usually begins at the shoot tip and proceeds downward. The shoot tip begins to twist, and further growth is stunted and distorted. Shoots may actually grow into knots. Flower clusters are occasionally attacked, resulting in undeveloped berries.

Control

Blackberry psyllid is only a problem when blackberries are grown in close proximity to conifers. Damage is avoided when blackberries are located 1.6 km or more from conifers, including any scattered ornamental plantings. Blackberry plantings located 0.4–0.8 km from pine woods have shown light infestations.

The traditional control method is to prune out affected canes after female psyllids die, since the nymphs are quite immobile. This method is effective if there are sufficient undamaged primocanes to provide the next season's crop. However, semierect thornless blackberries produce fewer primocanes than thorny blackberries, and cutting back stimulates lateral growth, making the canes unsuitable for trellising.

Insecticide applications are effective but must be carefully timed to reach the maximum number of migrating females before many eggs are deposited. Since the total female migration period occurs over about 14 days, the most efficient spray would probably be at 7 days after the first females are sighted. A second application could be required in 10 days if migration continues after the first spray.

Selected Reference

Peterson, A. 1923. The blackberry psyllid, *Trioza tripunctata* Fitch. N.J. Agric. Exp. Stn. Bull. 378. 32 pp.

(Prepared by L. C. Stuart)

Western Winter Moth

Occasionally larvae of the western winter moth, *Operophtera occidentalis* (Hulst), defoliate red raspberry sufficiently in the spring to pose an economic problem. This insect is generally considered a sporadic pest of red raspberries, particularly in the Pacific Northwest. At least three other species of winter moths, including *O. danbyi* (Hulst), *O. bruceata* (Hulst), and *O. brumata* (L.), may also occur on and defoliate *Rubus* spp. in North America.

Life History and Description

The general life history of winter moths begins with mated brachypterous females depositing eggs on the trunk and branches of deciduous trees in the late fall and winter. These eggs overwinter and hatch in the spring. The resulting larvae feed on blossoms, buds, and foliage through four or five instars. The larvae pupate in the soil during the summer and early fall months. Adults emerge in late fall and early winter. There is one generation per year.

Damage

Generally, damage has been most severe on red raspberry varieties and usually occurs on the field margins adjacent to hardwood trees. Borders of these fields should be inspected during spring regrowth for the presence of larvae and feeding damage.

Plant injury caused by winter moth larvae occurs early in the season. The most significant injury occurs as the larvae feed on developing floral buds and blossoms. Expanding leaves and floral parts are loosely webbed with silk, within which the larvae feed. Primordial floral and leaf tissues are destroyed by larvae boring through the buds. Petals, anthers, and ovarial tissues are usually consumed more extensively than leaf tissues.

In the Pacific Northwest an economic threshold in red raspberries of approximately eight larvae per hill has been used to indicate the need for insecticide application. Unlike the orange tortrix, larvae of the winter moths complete development and move into the soil to pupate prior to harvest. Larvae of the winter moth are not considered a contaminant of mechanically harvested fruit.

Control

If the action threshold is exceeded, a chemical or biological insecticide should be applied prior to bloom. Although larvae of the winter moth are quite susceptible to available registered insecticides, early application is recommended when the larvae are small and prior to the occurrence of extensive damage.

Few native natural enemies of winter moths have been reported from North America. General predators, such as lacewings, assassin bugs, and spiders, feed on the larvae. Larval parasitoids are generally uncommon in the species of *Operophtera*, but the most abundant are braconid wasps. The apparent absence of effective natural enemies suggests that regulation of winter moth populations may be primarily weather-related, as in the case of winter mortality.

Selected Reference

Miller, J. C., and Cronhardt, J. E. 1982. Life history and seasonal development of the western winter moth, *Operophtera occidentalis* (Lepidoptera: Geometridae) in Oregon. Can. Entomol. 114:629-636.

(Prepared by G. C. Fisher)

Raspberry Sawflies and Leaf Miners

The raspberry sawflies belong to the wasp family, Tenthredinidae. Of the several species that can be found associated with *Rubus* worldwide, the following have caused some economic damage.

The raspberry sawfly, *Monophadnoides geniculatus* (Hartig) in North America, is not a common pest; it occurs, at times, in low numbers. The small raspberry sawfly, *Priophorus morio* (Lepeletier), has been recorded in Europe and Australia, where it can be locally abundant in some years. The adults, which are similar, are small and thick-bodied and measure about 5-6 mm in length. The raspberry sawfly has one generation a year and lays eggs in leaf tissue in May and June; the small raspberry sawfly has at least two generations and prefers to oviposit in the leaf stalk from May until August. The small raspberry sawfly may also colonize blackberry, loganberry, *Sorbus* spp., and gooseberry. Larvae are light green, with many pairs of legs, and are marked by conspicuous bristles, which arise from small swellings on the body. Fully grown larvae are about 13 mm long. In the course of their development they consume leaf tissue, and heavy infestations may result in severe defoliation and the loss of the crop. Younger larvae generally feed on the outer edges of leaflets, but as they grow, they eat irregular holes throughout the leaf. In rare heavy infestations, only the larger veins are left uneaten. In North America, the larvae complete their feeding in less than 2 weeks and then estivate in cocoons, which they construct in the ground. Larvae remain in these cocoons during the winter and pupate in early spring. In the United Kingdom, the small raspberry sawfly may feed for longer periods and usually does not estivate, probably because of the lower temperatures. Pupation takes place in translucent, silken cocoons among the debris at the base of the plant or attached to leaves.

Two species of sawflies cause mines in *Rubus*: in North America, the blackberry leaf miner, *Metallus rubi* Forbes, and in the United Kingdom, the raspberry leaf-mining sawfly, *M. pumilus* (Klug). Both are small; adults of the former are about 5 mm long, and adults of the latter are about 3.5 mm long. The larvae of both species are white and 13 and 9 mm long, respectively, when fully grown. The first-generation larvae emerge from the eggs in June and begin to tunnel through leaf tissue and create "mines," causing affected areas to turn brown. When mature, they drop to the soil and pupate, usually in August. Second generation adults are active from mid-August to mid-September, when they lay eggs. After feeding, these larvae drop to the ground, where they overwinter. Pupation occurs in spring, and the cycle is repeated. Control is seldom necessary.

(Prepared by R. N. Williams)

Insects That Damage Canes

Rednecked Cane Borer

The rednecked cane borer, *Agrilus ruficollis* (Fabricius), is a serious pest of only commercial and wild *Rubus* spp. Its distribution in North America extends from Canada to the Gulf states and from New England to Minnesota.

Life History and Description
The life cycle of the rednecked cane borer has been described for several locations. In Arkansas, adults emerge from fruiting canes from early to mid-May until early June. They feed on the young foliage of primocanes for several days before laying eggs. The egg is cemented to the cane and hatches in 4–24 days, depending upon temperature. Larvae chew through the egg case and into the cane, where they tunnel in a spiral fashion around the cane. Generally three to four spirals under the bark result in the formation of a single gall. As the larva grows, it usually tunnels upward into the pith. However, tunnels may extend from 15 cm below to 64 cm above the gall. Overwintering larvae are best located by splitting the cane at the gall and following the brown, discolored tunnel from the gall to its end. Live larvae are rarely found inside galls.

In March, the larvae excavate an enlarged cavity in the cane. They become shortened and pupate by mid- to late April. The pupae molt to an inactive adult stage by early May. The adults require 10 or more days of maturation before they begin to chew through the cane wall, forming a D-shaped hole, as they emerge.

The rednecked cane borer adult is 6 mm long and all black, except for an iridescent coppery red to golden thorax (Plate 115). The white, legless larvae are 15–20 mm long and have a flattened head. The pupa is about the same length and shape as the adult but changes with age from creamy white to the color of the adult.

Damage
The feeding of the larva inside the primocane girdles the cane and causes a gall (Plate 116). Infested canes either die or become so weakened that they cannot support a full crop the following season. Girdling of the canes also predisposes them to winter injury.

Control
The best time to determine the presence and extent of rednecked cane borer damage is during winter pruning or after leaf fall. If the number of galled canes per row exceeds 10% of the primocanes or exceeds the expected number of primocanes to be pruned out per row, then chemical control of this insect is advisable. Primocane leaves should be inspected for adults twice weekly, beginning at first bloom. Insecticide is usually applied after first bloom and, if needed, repeated at weekly intervals, until no adults are observed on primocanes. Sprays are directed toward the base of the canes and are kept off of blooms in order to minimize bee poisoning.

Selected References

Chittenden, F. H. 1922. The red-necked cane borer. U.S. Dep. Agric. Farmers Bull. 1286:1-5.

Johnson, D. T., and Mayes, R. L. 1986. Biology and control of the rednecked cane borer. Arkansas Farm Res. 35:6.

Johnson, D. T., and Mayes, R. L. 1989. Biology and control of the rednecked cane borer, *Agrilus ruficollis* (F.) (Coleoptera: Buprestidae), on blackberries in Arkansas. J. Entomol. Sci. 24:204-208.

Riley, C. V., ed. 1870. The raspberry gouty gall. Am. Entomol. 2:103.

Slate, G. L., and Rankin, W. H. 1933. Raspberry growing in New York State. Cultural practices and disease control. N.Y. State Agric. Exp. Stn. Geneva Bull. 625. 62 pp.

(Prepared by D. T. Johnson)

Raspberry Cane Maggot

The raspberry cane maggot, *Pegomya rubivora* (Coquillett), also known as the raspberry maggot or the loganberry cane fly, is widely distributed on *Rubus* spp. throughout North America, continental Europe, Scandinavia, the Soviet Union, and the United Kingdom. The principal hosts are red raspberry, black raspberry, hybrid berries (loganberry, sunberry, and tayberry), blackberry (occasionally), European dewberry (*Rubus caesius* L.), some rose cultivars, and meadowsweet (*Filipendula ulmaria* (L.) Maxim.).

Life History, Description, and Damage
Adult flies, which resemble small houseflies, are dark gray and about 6 mm long. In the United Kingdom, they emerge in April and May and lay eggs in the axils of the terminal leaves of rapidly growing primocanes. After hatching, small, translucent larvae tunnel into the pith, where they burrow downward for a short distance before girdling the stem. This causes extensive damage to water-conducting tissues and causes the portion of the cane above the feeding site to rapidly wilt. This characteristic damage is known as limberneck in some parts of North America. The color of the wilted area may change. In tayberry, for example, the cane surface close to the site of damage rapidly changes from reddish green to bluish purple. Similar color changes have been reported in other hosts. Most wilted canes rapidly dry out and die, but occasionally the cane is not killed, and a swelling is produced at the site of girdling. Canes with such damage may continue to grow normally. The developing larvae continue to feed in the pith of the young cane. In red raspberry and some other *Rubus* spp. the larvae burrow to the base of the cane and pupate. The pupae remain in the cane in brown puparia until the following spring. Larvae infesting tayberry in Scotland only tunnel a short distance from the site of initial attack before pupating. Where this type of damage occurs, the cane produces secondary shoots from buds. The secondary shoots may also become infested, but only when attacks are severe. Branched canes are difficult to manage during autumn and winter cane pruning.

Control
Since damage to *Rubus* by the raspberry maggot is generally not severe, no specific insecticidal control measures have been developed. Growers with infested plantations must exercise a high level of sanitation by cutting out and destroying infested canes, preferably in the early summer, but definitely during winter pruning to reduce the fly population available to attack canes the following spring.

Selected References

Bovey, P., Tadic, M., and Pantellic, M. 1963. La mouch des pousses du framboisier (*Pegomyia rubivora* Coq.) en Suisse et en Yougoslavie. Rev. Pathol. Vég. Entomol. Agric. Fr. 42:71-76.

Gordon, S. C., and McKinlay, R. G. 1986. Loganberry cane fly—a new pest of *Rubus* in Scotland. Crop Res. 26:121-126.

Øydvin, J. 1976. Atak av bringebaerfluge *Pegomya rubivora* (Coq.) i hove til skottal og skothogd hos 22 familiar i bringebaer. Forsk. Fors. Landbruket 27:541-548.

(Prepared by S. C. Gordon)

Raspberry Cane Midge

The raspberry cane midge, *Resseliella theobaldi* (Barnes), and the associated disease complex (see Midge Blight) occur in many red raspberry growing areas of Europe but have not been recorded elsewhere.

Life History and Description

Adult raspberry cane midges are small (1.4–2.1 mm long), reddish brown, and, because of their similarity to other midges, are difficult to identify in the field. Midges overwinter as larvae in soil cocoons in the upper 1–4 cm of the soil. They pupate in the spring and emerge as adults from April until June, depending on the soil temperature. Males usually emerge first, and mating occurs soon after female emergence. Females oviposit in splits and wounds in the bark of primocanes. Suitable splits are usually found in the lower 50 cm of the primocane. In Scandinavia, splits in the bark of fruiting canes, especially in the raspberry cultivar Veten, are also used. Females oviposit by inserting their extended ovipositors under the flaps of wounds or splits to deposit small, elongated, translucent eggs.

Larvae hatch after 7–10 days and feed on the cortex of the cane (Plate 117). At first they are translucent, but they soon change to yellow or orange-pink. After about 14–21 days, fully grown larvae, measuring about 3 mm, fall to the soil surface to spin cocoons and pupate in the upper 1–4 cm. Adults of the second generation emerge in late June to August, usually at a time close to fruit harvest. By this time, many commercial red raspberry cultivars have developed extensive splits in the bark. Numerous eggs are laid in these splits at the base of the canes. The larvae feed on the newly exposed cane cork (periderm). Once fully grown, they drop to the soil and spin cocoons. In northern areas these larvae go into diapause, but in warmer regions they may produce a third generation.

Adult midges are weak fliers and are able to travel only short distances. Spread between distant plantations is most likely by the transport of cocoons attached to the roots of new planting material.

Damage

The direct damage caused by midge larval feeding is superficial, but the feeding sites soon become infected by a range of fungi, resulting in a disease called midge blight. These infections can lead to two types of damage. First-generation feeding sites develop deeply penetrating lesions, which give rise to conspicuous cankers (see Midge Blight). Although cankered canes may survive and fruit the following year, they are physically weakened and may be broken by the passage of machinery or fruit pickers or by winds during the late summer, autumn, and winter.

The damage arising from the second- and third-generation midges is more serious, because larger numbers of larvae are involved, and the fungi that colonize the larval feeding sites penetrate and damage the cork layer, which cannot be repaired at this stage of maturity. Characteristic irregular brown patch lesions develop during autumn and early winter (Plate 118). They are only visible when the outer layers of bark are scraped away. In comparison, the periderm of healthy vascular tissue is green. In severe outbreaks, the patch lesions may completely girdle the canes. In some older plantings, a spreading brown lesion caused by the pathogen *Leptosphaeria coniothyrium* (Fuckel) Sacc. has been found in association with midge-feeding lesions.

Affected canes continue to grow normally for the remainder of the first growing season without showing any visible symptoms. Because of their large basal diameter, these canes tend to be selected by the grower to fruit the following year. If patch lesions are extensive and girdling is common, a high proportion of these canes fail to produce lateral shoots, or they wilt and die before harvest. Yield losses of up to 90% have been recorded, but primocane growth is unaffected. The bark of most European red raspberry cultivars splits during the spring and early summer, making the plants highly susceptible to attack. The bark of European blackberries and some of the hybrid berries, such as tayberry and loganberry, does not split readily, and thus avoids midge damage.

Control

Control strategies for the raspberry cane midge are usually confined to the first generation, because the second and most important generation usually coincides with fruit harvest, limiting the use of insecticides. Since midges only oviposit in splits or wounds in primocanes, it is possible to interrupt the midge life cycle by manipulating primocane development in some cultivars. In the United Kingdom, the excessive vigor of the red raspberry cultivar Glen Clova has been controlled by applying dinoseb in oil (a desiccant herbicide) in the spring to "burn down" the first flush of primocanes. When this practice is used, the second flush of primocanes, which grow from buds at the base of the plant, are younger and almost free of natural growth splits. At this time the adult midges, which emerged in the spring, are ready to oviposit. Because of the lack of oviposition sites there is little or no first generation, and only a small and insignificant population survives to complete the second generation. The removal of primocanes by cutting produces the same beneficial effects, but this method is too labor-intensive and costly to be commercially viable. However, it may be worthwhile in smaller plantings.

Dinoseb in oil is no longer available in the United Kingdom, but alternative desiccants are being sought. Control of the cane midge is currently dependent on the use of insecticides, such as chlorpyrifos or fenitrothion. These are applied to the base of the canes when midges emerge in the spring. Two or three high-volume sprays (more than 1,000 L/ha) are directed at the basal 50 cm of the canes. Large volumes are required to induce runoff, so that the pesticide will flow down the cane surface and seep into the splits in the bark, killing the larvae. Timing of the sprays is difficult, because midge emergence is variable, even within a locality. Spray advisory systems (based on local scouting) or emergence forecasting (based on soil temperature measurements) are available in the United Kingdom.

Selected References

Gordon, S. C., Barrie, I. A., and Woodford, J. A. T. 1989. Predicting spring oviposition by raspberry cane midge from accumulated derived soil temperatures. Ann. Appl. Biol. 114:419-427.

Gordon, S. C., and Williamson, B. 1984. Raspberry cane blight and midge blight. Leafl. Minist. Agric. Fish. Food (G.B.). 905 pp.

Nijveldt, W. 1963. Biology, phenology and control of the raspberry cane midge, which is closely associated with fungal diseases or blight on raspberries in the Netherlands. Neth. J. Plant Pathol. 69:222-234.

Pitcher, R. S. 1952. Observations on the raspberry cane midge (*Thomasiniana theobaldi* Barnes). I. Biology. J. Hortic. Sci. 27:71-94.

Williamson, B., and Hargreaves, A. J. 1979. A technique for scoring midge blight of red raspberry, a disease complex caused by *Resseliella theobaldi* and associated fungi. Ann. Appl. Biol. 91:297-301.

Williamson, B., Lawson, H. M., Woodford, J. A. T., Hargreaves, A. J., Wiseman, J. S., and Gordon, S. C. 1979. Vigour control, an integrated approach to cane, pest and disease management in red raspberry (*Rubus idaeus*). Ann. Appl. Biol. 92:359-368.

Woodford, J. A. T., and Gordon, S. C. 1978. The history and distribution of raspberry cane midge (*Resseliella theobaldi* (Barnes) = *Thomasiniana theobaldi* Barnes), a new pest in Scotland. Hortic. Res. 17:87-97.

Woodford, J. A. T., and Gordon, S. C. 1986. Insecticides for the control of raspberry cane midge (*Resseliella theobaldi*) and midge blight. J. Hortic. Sci. 61:485-496.

(Prepared by S. C. Gordon and B. Williamson)

Raspberry Cane Borer

The raspberry cane borer, *Oberea bimaculata* (Olivier), belongs to the family Cerambycidae (the long-horned beetles) and is a native insect generally distributed over northeastern North America. Because of the similarity in their common names, there is some confusion about the raspberry cane borer and the rednecked cane borer. However, the damage they cause and the appearance of their adult insects are quite different.

Life History and Description

The adults of the cane borer are slender, dark beetles, about 1.25 cm long, with antennae as long as the body. The prothorax (area behind the head) is yellowish to bright orange, with two or three black dots on the upper surface. The beetles appear in June and may remain active until late August. After laying a single egg, the female girdles the cane about 10–15 mm above and 10–15 mm below the egg (Plate 119), which causes the tip of the shoot to wilt and die. Upon hatching, the larvae bore downward in the cane, where they overwinter not far below the point of the lower girdle. The next season, they continue to bore until they reach the crown. The second winter is passed at or below ground level. The following spring, full growth is attained, and the larvae pupate. New adults begin emerging in June; hence, 2 years are required to complete the life cycle.

Damage

Wilting shoot tips indicate the presence of the cane borer in a planting. Usually this is observed in early June. Twelve to 20 cm back from the tip, two girdles 2–3 cm apart confirm raspberry cane borer injury.

Control

The best cultural control practice is the destruction of the canes that show characteristic injury. If pruning is done within a few days after wilted tips appear, only 2.5 cm or so of the cane below the wilted portion needs to be removed. In order to be sure, the cut surface of the cane should be inspected, and if there is evidence of borer damage below the cut, short sections should continually be cut off, until all damaged tissue has been removed. All prunings should be removed and destroyed to kill the insect inside the cane.

Insecticides are not warranted in the control of the pest, since damage is seldom severe enough to cause concern.

Selected References

Johnson, D. T., and Schaefers, G. A. 1989. Bramble insects. Pages 73-85 in: Small Fruit Pest Management and Culture. University of Georgia Press, Athens.

Slate, G. L., Suit, R. F., and Mundinger, F. G. 1945. Raspberry growing in New York: Culture, diseases and insects. N.Y. Agric. Exp. Stn. Circ. 153:1-64.

(Prepared by R. N. Williams)

Tree Crickets

Tree crickets occur in many regions throughout the world. In North America there are about 16 species ranging from northern Mexico to southern Canada. They have numerous host plants, and several species have been found feeding on and depositing eggs in raspberry. Most injury to raspberry is caused by the blackhorned tree cricket, *Oecanthus nigricornis* Walker. The crickets are relatives of grasshoppers and, as indicated by the name, are arboreal in habit.

Life History and Description

Crickets undergo a single generation per year. They overwinter as eggs in the raspberry cane. Hatching occurs in May or June. The insects are considered beneficial by some, because they frequently feed on aphids, scales, or other soft-bodied insects. They seldom occur in sufficient numbers to have a major impact on these pests. The crickets also feed on foliage, floral parts, fruit, and various fungi. After undergoing five molts, they reach adulthood in August and deposit eggs through the end of September.

The bodies of the nymphs and young adults are semitransparent and pale whitish green and range from 6–25 mm long. They have dark antennae about twice as long as the body. Some specimens take on a blackish color later in the season. They have particularly large hind legs, which one normally associates with crickets.

Damage

Tree crickets are generally considered minor pests of raspberry. Nymphs and adults occasionally feed on the foliage and developing fruit, but most injury occurs because the crickets prefer to deposit eggs in the pithy raspberry stems (Plate 120). In this process, the adult first chews a small hole in the surface of the cane, and then inserts her ovipositor. Up to 80 eggs may be laid in a 25- to 75-mm row within the cane. These punctures tend to weaken the cane and may cause it to split or break under fruit load or wind stress. Punctures lower down on the cane may inhibit fruit development. These punctures may also serve as entry points for disease. Up to 75% of the canes in some *Rubus* plantings have been observed to contain punctures.

Control

Natural control occurs through the attacks of digger wasps and by about eight species of egg parasites.

Cultural control is achieved through clean cultivation, including the elimination of excessive weed growth. The elimination of wild brambles for the control of other insects and pests will also help reduce tree cricket populations. The most practical method of control is pruning out and burning the punctured canes containing eggs during normal pruning operations in the spring. The mowing-off and destruction of canes in annual cropping practices would be equally effective.

Chemical control is seldom required, but when used it should be timed to kill the adult females prior to oviposition during August.

Selected Reference

Fulton, B. B. 1915. The tree crickets of New York: Life history and bionomics. N.Y. State Agric. Exp. Stn. Tech. Bull. 42. 47 pp.

(Prepared by G. A. Schaefers)

Rose Scale

The rose scale, *Aulacaspis rosae* (Bouché), is an armored scale insect in the family Diaspididae that attacks the canes of *Rubus* spp. and the stems of roses growing in humid, shady places. It is distributed worldwide wherever *Rubus* spp. and roses are grown.

Life History, Description, and Damage

In most of North America, there are at least two generations annually. The first begins with egg hatch on the canes in late May or June, and a second begins in August. First-stage crawlers walk around on the canes for a day or two before settling down to feed. Once feeding begins, they become sedentary and secrete a hard, shell-like covering, or scale. They insert their threadlike mouthparts into the epidermis of the cane and feed on plant sap. Once mature, the adult males, which are not sedentary, actively seek out the sedentary females and mate. After mating, adult females remain sedentary and continue to feed and produce eggs.

Female scales are round, ranging from 2–3 mm in diameter. The shell is dirty white, with an orange-yellow dot in the center. This scale covers orange or pinkish bodies and red eggs. The males are considerably smaller, long and narrow, and snow white. Adult females are wingless; males are winged and able to fly.

Scale infestations on *Rubus* appear as small, white, circular structures on the canes (Plate 121). These scales generally occur at the base of the cane but may cover much of its surface. In severe infestations, the scales may overlap and completely cover and encrust the bark, causing the cane to look whitewashed. In addition to affecting the appearance of the plant, scale insects can also reduce plant vigor. Serious infestations may result in reduced growth; irregular fruit ripening; small, poor-quality fruit; and, in extreme cases, death of canes.

Control

Sanitation is important to reduce scale populations. Infested canes should be removed and destroyed to suppress further buildup and spread. The elimination of wild brambles near cultivated plantings may also be beneficial. The rose scale is attacked by several predators and parasites, which can provide some level of natural control. It is not usually a serious pest in commercial *Rubus* plantings, as it is quite susceptible to most insecticides that are applied for control of other pests. Liquid lime sulfur is also effective in the control of rose scale and is commonly used as an early-season (dormant) fungicide for the control of several diseases.

Selected Reference

Smith, J. B. 1902. The rose scale, *Diaspis rosae* Bouché. N.J. Agric. Exp. Stn. Bull. 251:1-14.

(Prepared by R. N. Williams)

Stalk Borer

The stalk borer, *Papaipema nebris* (Guenée), belongs to the family Noctuidae. This insect is almost a universal plant feeder, attacking stems of any plant large enough to shelter it, and soft enough that it can bore into the stem. The primary symptoms in a raspberry field are the wilted tips and large holes in the side of the cane, 6–8 inches back from the tip. Damage is done primarily by boring and tunneling in the canes. Plantings adjacent to weedy areas containing a favored host, such as giant ragweed, are most often attacked. Specific control measures are generally not warranted.

(Prepared by R. N. Williams)

Insect Contaminants of Mechanically Harvested Fruit

About one half to two thirds of the red raspberries grown in the Pacific Northwest and a negligible portion of those produced in Europe are machine harvested. Machines are labor-saving, but they remove many insects and spiders along with ripe berries. These contaminants can be difficult and expensive to remove at the processing plant. If not removed, they present legal and economic difficulties to the processor because of Food and Drug Administration standards for insect parts in processed raspberries and from consumer resistance to the presence of such material in the berry products they buy. In extreme cases, growers may have their berries reduced in price or even rejected by the processor because of too many insects in harvested fruit.

Over 60 families of insects, as well as spiders, have been identified as living on raspberry foliage. Most of these are either beneficial as predators or have no effect on raspberry plants. However, all are potential contaminants of the harvested product. Some of the most important insects and those that are difficult to control are adult root weevils and cutworms, including the orange tortrix larvae, stinkbugs, and miscellaneous lepidopteran larvae. Other insects that have caused problems for growers are the large raspberry aphid and *Lygus* spp. Spiders can also be a problem. In experimental machine harvester trials in Scotland, the European (or common) earwig, *Forficula auricularia* L., was frequently present in the fruit samples.

Control

Properly designed and operated air cleaners on the harvesters will remove most of the small and light-bodied insects and spiders. Heavier insects, such as caterpillars and large beetles, might not be removed by the air cleaner and must be removed by hand from the sorting belt. These insects should be dropped into a pail of water containing a small amount of detergent to break surface tension, thereby killing them. They should not be discarded back into the field, where they could reinfest the raspberry plants.

In spite of air cleaners, unusually high populations of small insects, such as aphids, gnats, and Lygus bugs, may result in considerable contamination of harvested berries. Therefore, it is often desirable to apply a broad-spectrum, short-residual insecticide as a clean-up spray before harvest begins.

Selected References

Antonelli, A. L., Shanks, C. H., Jr., and Fisher, G. C. 1988. Small Fruit Pests. Biology, Diagnosis and Management. Wash. State Univ. Coop. Ext. Serv. Ext. Bull. 1388. 20 pp.

Kieffer, J. N., Shanks, C. H., Jr., and Turner, W. J. 1983. Populations and control of insects and spiders contaminating mechanically harvested red raspberries in Washington and Oregon. J. Econ. Entomol. 76:649-653.

(Prepared by C. H. Shanks, Jr.)

Part III. Disorders Caused by Abiotic Factors

Nutritional Disorders

Plant nutrition can be affected by temperature, light, and crop load and the moisture, pH, type, and nutrient content of soil. Plants may show deficiencies in one or more nutrients under extreme edaphic or climatic conditions. Deficiencies can reduce growth, yield, and fruit quality and can make plants more susceptible to abiotic and biotic diseases.

Nutrient deficiencies can be difficult to visually diagnose. First symptoms are often a general reduction in plant growth or vigor. Distinct foliar symptoms usually develop only when deficiencies become extreme. Some deficiency symptoms can easily be confused with those caused by viruses and other pathogens or with abiotic disorders, such as herbicide or low-temperature injury. In addition, it is rare that a deficiency problem is caused by only a single nutrient. For example, when soil pH is too high, manganese, iron, and zinc may become deficient. When soil pH is too low, calcium, phosphorus, and magnesium may become unavailable, and aluminum and manganese may become toxic. Thus, a reliable diagnosis is difficult to make by visual inspection.

Analytical tests can be used to help the grower maintain proper nutrient status. Soil tests prior to planting can aid growers in adding necessary nutrients and adjusting soil pH. Both soil and foliar analyses are useful in established plantings to determine fertilization practices and requirements. Many governmental and private organizations offer assistance in collecting samples and interpreting results for specific site conditions. Little information exists on nutrient responses in blackberries; however, deficiency symptoms and the physiological role of nutrients should be similar among the *Rubus* spp.

Nitrogen

Nitrogen makes up 2–4% of a plant's dry matter. It is an essential constituent of numerous organic compounds, such as amino acids, proteins, and nucleic acids. Nitrogen status is visually determined by evaluating growth and leaf color. Its deficiency is characterized by short, weak primocanes, short internodes, small leaves, and premature leaf abscission. Nitrogen is mobile in the plant, so foliar deficiency symptoms (chlorosis or yellowing) first show in the older leaves. Necrosis of leaves or parts of a leaf can eventually occur if the deficiency is severe. Some studies have shown an increase in raspberry yield (mainly from an increase in primocane and berry size) in response to nitrogen application. However, others have found no response. Nitrogen toxicity is rare, but excessive nitrogen may be indicated by high plant vigor, long internodes, dark blue-green leaves, excessive suckering, low yield, and poor quality. Brambles high in nitrogen may also be more susceptible to winter injury, and imbalances with other nutrients can be affected.

Phosphorous

Phosphorus is required for many metabolic or energy-requiring processes in the *Rubus* plant, including root growth and ripening of berries and seeds. It makes up 0.15–0.30% of a leaf's dry weight. Phosphorous deficiency is uncommon in most growing regions if the pH is in the optimal range (about 5.8–6.8 for *Rubus* in Oregon). Plants suffering from this deficiency are usually retarded in growth. Symptoms appear first in the older leaves, which are often darkish green with red or black areas. Older leaves may fall prematurely. Root growth may be retarded in deficient plants, and yields may be low, because of poor-quality fruit (poor fruit set). In some plants phosphorus nutrition has been related to flower bud initiation or development. Excess amounts can depress growth and may induce zinc, iron, and copper deficiencies. Few specifics are known about phosphorus nutrition in *Rubus*.

Potassium

Potassium is required in large quantities by the bramble plant and makes up 1–2% of the leaf dry matter. It plays an important role in the water status of the plant, enhances the movement of photosynthates (food), and activates various enzyme systems. The need for it is greatest during fruiting. Potassium deficiency may be more common in heavy cropping years, in acid soil, in periods of drought, in waterlogged or wet soils, or in sandy, organic, or calcareous soils. It does not immediately result in visible symptoms. First, there is a reduction in growth rate, with chlorosis and necrosis of leaves occurring later. In the older leaves, marginal and tip chlorosis and necrosis occur and progress inward, toward the midrib. Basal leaf scorch may occur. Leaves may curl and wilt easily. Other crops deficient in potassium have exhibited decreased resistance to drought and increased susceptibility to frost damage and fungal attack.

Sulfur

Sulfur is an essential constituent of proteins. Its content in plant tissues is between 0.2 and 0.5% dry weight. Symptoms of its deficiency are similar to those of nitrogen deficiency. However, unlike nitrogen, sulfur is not mobile in the plant; therefore, chlorosis occurs first in the young leaves.

Magnesium

Magnesium is an essential constituent of chlorophyll, the green pigment in leaves, which is involved in the manufacture of carbohydrates. It is also important in many biochemical reactions and enzyme activation. Magnesium deficiency is more common on light, acid soils with low magnesium content, sandy soils with a relatively high potassium content, and calcareous soils. Magnesium is mobile in the plant, and deficiency begins in the older leaves first. Interveinal chlorosis occurs, and in extreme cases the areas become necrotic. Symptoms may be confused with those caused by viruses (yellowing) or potassium deficiency. Some leaves, particularly those exposed to sunlight, may have a withered appearance. Generally they are stiff and brittle, with twisted veins. Magnesium-deficient leaves often fall prematurely.

Calcium

Calcium is involved in many physiological processes and is a component of plant cell membranes. It makes up between 0.6 and 1.5% of a leaf's dry weight. Calcium deficiency is rare in brambles grown in soils with an optimal pH. Factors that inhibit transpiration, such as high humidity or drought stress, can lead to this deficiency in fruit. Also, calcium uptake by roots may be inhibited by poor aeration or by low soil temperatures, which induces a deficiency.

Iron

The average content of iron in leaf tissue varies between 30 and 100 ppm dry weight. Iron, like magnesium, is required for chlorophyll production. However, unlike magnesium, iron deficiency begins in the younger leaves and is characterized by interveinal chlorosis. It is most common when soil pH exceeds 7.5.

Manganese

Manganese activates many enzymatic processes and is involved in photosynthesis. Manganese deficiency is most common on alkaline, sandy soils high in organic matter. Symptoms resemble those of magnesium deficiency, since in both cases interveinal chlorosis occurs in the leaves. In contrast to magnesium deficiency, in which the older leaves are first affected, manganese deficiency symptoms are first visible in the younger leaves. The first symptoms are small, yellow spots on the leaves. Later, the entire interveinal area may become chlorotic. This deficiency is rare in brambles, unless the soil pH is above 7.5. The content of manganese in leaf tissue may vary widely, depending upon soil pH (30–500 ppm dry weight). A positive effect of manganese on the vitamin C content of raspberry fruit has been documented.

Zinc

Zinc is required for the activation of many enzyme systems, for nitrogen metabolism, and for the synthesis of auxin (a growth hormone). The average content of zinc in leaf tissue is between 30 and 50 ppm dry weight. Zinc availability may be reduced in soils with high phosphorus availability, because an insoluble precipitate is formed. Plants suffering from this deficiency often show interveinal chlorosis of the leaf. These areas are pale green, yellow, or even white. Leaves may have a green margin and be contorted. Since zinc deficiency impairs auxin synthesis, shoot growth is retarded. Affected shoots produce small, stiff leaves and have a rosetted appearance.

Copper

Copper plays an important role in many biochemical processes, including photosynthesis. The average content of copper in leaf tissue is between 5 and 15 ppm dry weight. Deficiency symptoms may include interveinal chlorosis and a green margin on young leaves, poor shoot growth, and abnormal fruit development. Symptoms have not been documented for brambles in field situations.

Molybdenum

Molybdenum is required in very small quantities by plants; therefore, deficiencies are rare and have not been documented for brambles in field situations. Deficiency symptoms may be characterized by a gray necrosis and upward folding of the edges of young leaves. As leaves age, necrosis spreads toward the midrib.

Boron

Boron deficiency affects the level of auxin and cytokinin (growth hormones) in plant tissues. It is also involved in pollen tube growth and thus affects fruit set. The average content of boron in leaf tissue is between 20 and 50 ppm dry weight. Boron deficiency symptoms include abnormal or uneven shoot growth in the spring; misshapen, dark blue-green young leaves; abnormally shaped fruit; and underdeveloped (stubby) roots. This deficiency may also delay fruit ripening. Excess boron is toxic to plants. Therefore, leaf tissue levels should be monitored to judge boron requirements.

Selected References

Chaplin, M. H., and Martin, L. W. 1980. The effect of nitrogen and boron fertilizer applications on leaf levels, yield and fruit size of the red raspberry. Commun. Soil Sci. Plant Anal. 11:547-556.

Dilli, R. B. 1983. Role of nitrogen fertilization on foliar composition, growth, and yield of 'Heritage' red raspberry (*Rubus idaeus* L.). M.S. thesis, Ohio State University, Columbus.

Harris, G. H. 1943. The effect of micro-elements on the red raspberry in coastal British Columbia. J. Amer. Soc. Hortic. Sci. 45:300-302.

Kowalenko, C. G. 1981. Effects of magnesium and potassium soil applications on yields and leaf nutrient concentrations of red raspberries and on soil analyses. Commun. Soil Sci. Plant Anal. 12:795-809.

Kowalenko, C. G. 1981. Response of raspberries to soil nitrogen and boron applications. Commun. Soil Sci. Plant Anal. 12:1151-1162.

Kowalenko, C. G. 1981. The effect of nitrogen and boron soil applications on raspberry leaf N, B, and Mn concentrations and on selected soil analyses. Commun. Soil Sci Plant Anal. 12:1163-1179.

Mengel, K., and Kirby, E. A. 1982. Principles of Plant Nutrition. 3rd ed. International Potash Institute, Bern, Switzerland. 665 pp.

Pritts, M. 1989. Nutrient management. Pages 85-93 in: Bramble Production Guide. M. Pritts and D. Handley, eds. Northeast Regional Agricultural Engineering Service, Ithaca, NY.

Tomkins, J. P., and Boynton, D. 1959. A response to potassium by the black raspberry. J. Amer. Soc. Hortic. Sci. 73:164-167.

(Prepared by K. Wilder and B. C. Strik)

Herbicide Injury

Minimizing the impact of weeds in a raspberry or blackberry planting is critical to ensure the production of high-quality fruit. Weeds not only compete directly with the crop for water, light, and essential nutrients but also can harbor harmful insects, create an environment favorable for the development of diseases, interfere with harvesting operations, and affect the aesthetic quality of a bramble planting.

Crop rotation, cover cropping, fumigation, mulching, mowing, and cultivating are components of the weed management strategy. Another important component is chemical weed control. Chemical control dramatically increases the efficiency of production, since labor inputs are greatly reduced. Herbicides can also reduce the energy required for mowing or cultivating and, in turn, lessen soil compaction. Cultivating young plantings may also injure the root system, causing reduced plant growth and poor primocane emergence in the following year. In many cases, herbicide-treated plantings perform better than those for which cultivation is used for weed control.

Herbicides must be applied carefully, to ensure that an amount sufficient to effectively control weeds is applied at levels below those which may injure the bramble plant. Most herbicides can be toxic to *Rubus* if applied improperly or at excessive rates, but injury seldom occurs in the field, unless unusual weather conditions prevail. Most cases of herbicide overdose result from erroneous sprayer calibration, spray overlap, drift, application at the improper stage of plant growth, or accumulation or carryover in the soil.

Diagnosing Herbicide Injury

Although herbicide injury is rarely observed in the field, it is useful to identify it when it does occur, so that corrective measures can be taken. It is expressed in many different ways, and symptoms are often similar to those caused by nutrient deficiencies, salt burn, or certain diseases. Patterns of herbicide injury may follow changes in soil type, drainage, and previous cropping designs, or injury may appear on one cultivar but not another. Injury caused by some biological agents or nutrient imbalances may tend to follow similar patterns, so it may be difficult to identify the causal factor. In most cases, however, herbicide injury follows regular patterns in the field. These patterns could be caused by spray overlap (indicated by narrow bands of injury in the direction of the rows), by the sprayer slowing down to make turns (indicated by injury at the end of every other row), or by improper amounts of herbicide mixed in the spray tank (indicated by the abrupt termination of injury symptoms when the sprayer is refilled). It is also important to note the time between the observation of injury and the application of herbicides. A specific herbicide can often be discounted as a cause of injury when the interval between application and observation of injury is not appropriate.

Herbicide injury symptoms may differ, depending on whether application was made to leaves or soil. For example, if a herbicide is applied at high rates or is missapplied directly to leaves, the leaves may become scorched or necrotic very soon after the application. Symptoms such as interveinal yellowing and necrosis may develop over a period of several weeks. These types of chronic symptoms are more typical of soil-applied herbicides, which are absorbed by the roots and transported through the vascular system of the plant. In many cases, herbicides simply cause a reduction in growth without visible symptoms of foliar injury.

If herbicide injury is suspected, then the fruit should not be eaten or sold, since unacceptably high residues could be present. Growers should take all possible precautions to avoid herbicide misapplications in "pick-your-own" plantings, since customers may not recognize foliar symptoms of injury.

Differential Sensitivity

Not all plants are equally sensitive to herbicide injury. For example, newly transplanted raspberry and blackberry plants are more susceptible than established plants. Many labels recommend that growers wait to apply the herbicide until soil has settled around the roots after transplanting or until the second year for application. Others recommend that lower rates be used during the establishment year.

Plants propagated directly from tissue culture are more sensitive to preemergent herbicides than are rooted cuttings or suckers. Reductions in plant growth are often observed if certain herbicides are applied to tissue-cultured plants within 8 weeks of planting. Some cultivars are also more susceptible to injury than others. Purple raspberries tend to be more sensitive than either black or red raspberries, and differences exist among cultivars of the same species.

Chemical Properties and Modes of Action

Injury symptoms are caused by the specific chemical properties of the herbicide, which allow the chemical to interfere with critical life-sustaining processes in the plant. Most herbicides are intended to be selective but cause injury when these selective mechanisms break down. Many herbicides bind tightly to soil particles, so they remain near the soil surface. Weed seeds germinating near the soil surface are killed, but the raspberry roots growing several inches below the surface are unaffected. Herbicides with this chemical property may cause injury if applied to coarse-textured (sandy) soils or at high rates or used repeatedly from year to year. Those that do not bind as tightly to soil particles or are quite soluble in water may injure plants following periods of heavy rain.

Some herbicides can come in direct contact with the leaves of raspberry and blackberry plants and not injure them. They may be rapidly degraded by the plant, or they may affect a biochemical pathway that does not exist in a raspberry plant, but does exist in a weed species. Some herbicides that could cause injury to *Rubus* spp. are applied during the dormant season, when no leaf uptake can occur, or they are directed toward the base of the canes and are not readily transported to the leaves.

Herbicides affect plants through many modes of action, including inhibition of cell division, mitosis, photosynthesis, metabolism, or wax formation on leaves; interference with xylem differentiation; stimulation of abnormal growth; or destruction of the cell membrane. By knowing how the herbicide works (or its chemical classification), whether the application was made to roots or leaves, and how the herbicide is translocated in the plant, one can predict how injury symptoms might be expressed.

Only a few of the more than 150 phytotoxically active compounds are discussed in this section. They are from various chemical classifications and have a history of use in *Rubus* plantings in North America. In general, herbicides of the same chemical classification cause similar injury symptoms.

Preemergent Herbicides

Simazine (Triazine)

Simazine works primarily on broadleaf weed seedlings by inhibiting photosynthesis. It can cause reductions of cane growth, particularly on young tissue-cultured plants and purple raspberries. When visual symptoms appear, it is likely that application rates were too high, that soil accumulation occurred through repeated use on a low-pH soil, or that tender foliage was exposed to the simazine spray. Symptoms generally appear as an initial yellowing of the leaf margins and interveinal areas (Plate 122). Yellow margins may then turn brown, giving the leaf a scorched appearance. Leaf scorching may later spread to the interveinal areas. In severe cases, only the main veins remain green. Eventually the entire leaf may turn brown and die. Certain cultivars may express differential symptoms when injury is minor and chronic (for example, veinal yellowing in Royalty purple raspberry).

Terbacil (Uracil)

Terbacil, like other uracil herbicides, is an inhibitor of photosynthesis. Because of its ability to burn exposed leaves, applications are not recommended on actively growing plants or on first-year plantings. Injury occurs when rates have been excessive, or when accumulation occurs from repeated use. With severe terbacil injury, leaves turn yellow in the interveinal areas, and eventually these areas become necrotic (Plate 123). Leaf necrosis may also occur without significant chlorosis (Plate 124). Occasionally, only the veinal areas of the leaf will remain after the interveinal areas have died. When injury is mild and chronic, veinal yellowing can occur without noticeable interveinal yellowing. Plants injured by terbacil are very susceptible to winter injury and rarely recover if leaf necrosis has occurred. Terbacil injury is most common on sandy soils, on those low in organic matter, or where water tends to puddle on the soil.

Dichlobenil (Nitrile)

Dichlobenil is a strong inhibitor of growth, seed germination, and actively dividing cells of meristems. It acts primarily on growing points at root tips and in buds and is very effective against a broad spectrum of germinating weed seedlings. The first symptom of injury in *Rubus* is a reduction in growth, followed by the appearance of leaves that are darker than normal. Leaf margins begin to turn yellow, and then brown. Injury occurs most often when application of the granular formulation is not uniform. Dichlobenil does not accumulate in low-organic matter soils with repeated use, since it is very volatile. Injury may occur if applications are made on young plantings, on a hot day, or in extremely sandy soil. *Rubus* plants usually avoid injury, because of their deep root systems.

Diuron (Urea)

Diuron and other urea herbicides block a critical step in the photosynthetic pathway of plants. Acute symptoms occur when high concentrations are applied directly to the leaf. Within a few days, light green areas appear, which take on a water-soaked appearance and then become necrotic. Chronic symptoms occur with excessive root uptake. This occurs when high rates are used on coarse-textured soils or those low in organic matter. The most prominent symptom is the yellowing of leaf veins, with the interveinal areas remaining green. Symptoms are accentuated in cool weather. Selectivity in *Rubus* exists because the diuron has low water solubility and high adsorptive properties in loam and clay soils. It will control shallow-rooted annual weeds without affecting the more deeply rooted raspberry or blackberry crop, since it does not usually leach more than one inch in most soils.

Norflurazon (Fluorinated Pyridazinone)

Norflurazon is a preemergent herbicide that markedly reduces chlorophyll production in most plant species. Injury symptoms in *Rubus* are very distinctive. The veins of lower leaves first become bleached, and the interveinal areas remain starkly green (Plate 125). The bleaching slowly progresses into the interveinal areas, following the secondary veins. Eventually leaf margins begin to brown. Although norflurazon binds to soil particles and does not leach readily, injury has occurred where soil moved from the site of application and accumulated through erosion. Injury is also common where norflurazon is used in consecutive years, because it has a longevity in soils of up to 15 months. The likelihood of foliar injury from norflurazon is greater than that from the other preemergent herbicides.

Oryzalin (Dinitroaniline)

Oryzalin inhibits growth of the entire plant primarily by limiting production of lateral roots. It is very effective on small seedlings that have germinated in the top few centimeters of soil. It rarely affects *Rubus* plants, except where application was made to soil containing cracks, or where heavy rains followed an application to sandy soil. Distinctive foliar patterns are usually not apparent with oryzalin injury, except when sprays contact young, tender foliage, which would occur during late spring applications or in treatments of tissue-cultured plantlets. Injury symptoms include stunted leaves with puckered interveinal areas, accompanied by a mottled yellowing. Mature plants will outgrow this injury, but young, tissue-cultured plants generally do not recover. Oryzalin is bound tightly by organic matter, so it is not recommended for use on high-organic matter soils.

Napropamide (Amide)

Napropamide inhibits root growth, particularly that of grasses. It also inhibits the sprouting tubers of nutsedge. Because its mode of action is relatively selective, it is very safe to use in *Rubus* plantings, and injury symptoms are rarely observed in the field, even with tissue-cultured plants. Marginal leaf chlorosis and growth reduction can occur from severe over-application.

Postemergent Herbicides

Translocatable Herbicides

Glyphosate and Other Translocatables

Glyphosate is a nonselective, translocatable, postemergent herbicide that controls most annual and perennial weeds by inhibiting critical biochemical pathways. It is metabolized very slowly in the plant. Glyphosate is used in *Rubus* plantings as a directed spray to avoid contact with foliage. It can translocate from suckers in the alleyway into the parent plant. It is tightly bound by most soils, so it does not move into the root zone or accumulate from repeated use. Injury commonly occurs when the *Rubus* leaf or green cane comes in contact with the herbicide during application or drift. Symptoms may take several weeks to develop, especially with cool, cloudy weather. In *Rubus*, the leaves on the tops of primocanes turn very light yellow (Plate 126), and the newly formed leaves do not fully expand. Eventually these leaves turn brown and die (Plate 127). Symptoms are also expressed the following year when the plant produces straplike leaves (Plate 128). Glyphosate is most damaging to *Rubus* when applied in late summer or early autumn, because translocation is then occurring downward, into the root system.

Translocatable herbicides such as sethoxydim and fluazifop are very safe to use in *Rubus* plantings, since only grasses are injured by these chemicals. Slight marginal burning has occasionally been reported by growers after application, but this is due to the added surfactant in combination with spraying under hot, humid conditions.

Contact Herbicides

Contact herbicides were developed to rapidly kill any foliage to which spray is applied; therefore, visible injury symptoms develop quickly, often within 24 hr. Most do not have residual soil activity. They are used in *Rubus* plantings as directed sprays, to burn off annual and perennial weeds without injuring the canes. Some are being considered for use in suppressing primocanes. Injury results when application is made to *Rubus* foliage, since these herbicides are not selective among the broad-leaved plants.

Paraquat (Bipyridilium)

Paraquat is a nonselective herbicide with fast contact action. It is biologically inactive in most soils, so root uptake does not occur. However, foliar injury symptoms will appear within 24 hr of application to leaves. The waxy epidermis of the lower portion of *Rubus* canes permits tolerance of paraquat sprays in early spring before budbreak. Application after leaf emergence results in rapid browning, upward curling, and desiccation of leaves where contact with spray has occurred (Plate 129). Light accelerates the development of phytotoxic symptoms.

Diphenyl Ethers

Diphenyl ethers, such as oxyfluorfen, are considered to be contact herbicides, but they do have limited preemergent activity. They can be absorbed by *Rubus* leaves, but little long-distance transport occurs within the plant. Browning and desiccation of sprayed leaves and growing points occur within days after application, and symptoms are accentuated under high-light conditions. Oxyfluorfen is very resistant to leaching, so injury to the *Rubus* plant will generally not occur through root uptake.

Glufosinate

Glufosinate is a phosphate compound used as a contact herbicide. It has some systemic activity but no soil activity.

Injury to *Rubus* occurs through the slow desiccation of leaves. Signs of leaf browning occur within 5 days of application, and this may be accelerated in warm weather. Browning of leaves and stunting of primocanes will occur when application occurs in early summer, but generally the canes are not killed if they are more than 10 cm tall. The tops of primocanes will later become bushy, caused by the breaking of axillary buds. Leaves on affected canes will be thick, small, and not completely expanded and exhibit yellowish margins (Plate 130).

Volatile Herbicides

Injury has been reported from herbicides that drift from the area of application into *Rubus* plantings. These are most often volatile materials, such as 2,4-D (a phenoxy compound) or dicamba (a benzoic compound). Herbicide injury from drift usually decreases in a regular pattern in the direction of the prevailing wind at the time of application. Injury symptoms from 2,4-D and dicamba appear as curling of the tops of the primocanes, with an upward folding of leaves and browning of leaf margins (Plates 131 and 132). Injury can be very severe in tissue-cultured plantlets. Severe injury can result in abnormal growth and malformation of the leaves.

Correcting Herbicide Excesses

If excessive residual herbicides are suspected in a site, as would be the case where atrazine was recently used to grow corn, one should test for herbicide carryover by performing a bioassay. This test involves planting sensitive vegetable or grain seeds in a sample of soil from the site. Specific vegetable or grain species are recommended for the herbicide in question. If the sensitive plant exhibits injury symptoms, then corrective measures are warranted.

Waiting for a period of time before planting the field with raspberries or blackberries is one alternative, if herbicide carryover is high. Excessive herbicides will eventually dissipate through photodecomposition, chemical decomposition, biological degradation by microorganisms, detoxification by weeds, volatilization, adsorption by organic matter and clays, or leaching and runoff. The time required for an excessive amount of herbicide to be reduced to a safe level depends on many factors, including the herbicide's chemical properties and amount in the soil, temperature, rainfall, and plant community in the site. A bioassay should be performed during late fall or winter to assess the suitability of a site for planting the following spring.

A second alternative is deep plowing to dilute the herbicide. This works well for herbicides that are relatively nonmobile in the soil. Third, if herbicides are in coarse-textured, low-organic matter soils, leaching with sprinkler irrigation may be effective. A fourth alternative is treating the problem area with activated charcoal. Certain herbicides will adsorb to the surface of the charcoal and lessen their effects on the plant. Activated charcoal is often used where excessive rates of herbicides have recently been applied to established plantings. The amount of charcoal required to inactivate 1 kg of herbicide will depend on the herbicide, the rate of application, soil properties, and the quality of the charcoal.

With raspberries and blackberries, the long-term effects of translocatable herbicides, such as glyphosate, can be reduced if all canes are removed in late summer, after the herbicide has been applied. Some of the herbicide will be removed with the canes, leaving less to be transported into the crown as autumn approaches. Plants treated in this manner will have a greater chance of recovery than those with retained canes.

(Prepared by M. P. Pritts)

Environmental Stress

The adverse effects of environmental stress are often difficult to recognize. The symptomatology of environmental stress in *Rubus* spp. is not well defined, except in extreme situations, such as drought (wilting); tissue death due to freezing, desiccation, or high-temperature injury (foliar or vascular necrosis); or growing plants in soil that falls below the required nutrient levels to sustain normal growth (nutrient deficiency symptoms). In addition, it is often difficult to separate the adverse effects of environmental stress from the effects of other abiotic disorders, insects, and diseases. Each factor or stress interacts with other environmental factors as well as with a plant's specific physiological condition at any given moment. For example, if a young, succulent raspberry cane is exposed to a temperature of $-7°C$ in the spring, it will likely die. The same cane in late fall, after a period of winter acclimation, may be unaffected. If other stresses such as wet soils, nutrient deficiencies, diseases, and insect feeding are present, the cane may be more susceptible to low-temperature injury, even when fully dormant.

The effects of environmental stress, such as cold-temperature (winter) injury, undoubtedly play a role in the development of certain diseases by predisposing the plant to infection. Although it seems to be common knowledge that this relationship exits, there is very little quantitative data in the literature to support this general assumption for *Rubus* spp. Nevertheless, the adverse effects of environmental stress need to be seriously considered in commercial production, since they can have pronounced effects on the pest management strategies used by growers as well as on the profitability of the operation.

Temperature

Low-Temperature Injury (Winter Injury)

Cold-temperature injury to *Rubus* spp. can occur in several ways. Late frosts in the spring or early summer can result in severe flower mortality and complete loss of the crop in summer-bearing *Rubus* spp. Early fall frosts or freezes may prematurely terminate fruit production and plant growth of primocane-bearing types of raspberries and result in significant yield reductions. If cold injury is prevalent in a specific region or planting, the grower may choose to select later-bearing cultivars of summer-bearing raspberries or blackberries and earlier primocane-bearing cultivars of red raspberry to avoid this problem.

The most prevalent low temperature–related injury generally results from extremely low or fluctuating temperatures during the dormant season. *Rubus* spp. and cultivars vary greatly in their tolerance to cold temperatures. The hardier red raspberries generally tolerate temperatures to about $-29°C$, with purple raspberries tolerating them to $-23°C$; black raspberries, $-20°C$; erect blackberries, $-18°C$; and trailing blackberries, -13 to $-11°C$. It must be emphasized that these temperatures are only estimates and represent cold-tolerance levels for the hardier species and cultivars. In addition, the cold tolerance of each cultivar or plant is directly affected by several other environmental factors.

Most *Rubus* spp. enter a rest, or dormant, period between their first and second growing seasons. Exposure to low temperatures (chilling) removes dormancy. Raspberries require

800–1,600 hr of chilling at temperatures below 7°C to satisfy their rest requirement and remove dormancy. Blackberries generally require 300–600 hours of chilling. Plants are most tolerant to low temperatures when they are fully dormant. Most winter injury probably occurs prior to or after the plant is fully dormant.

When temperatures rise after plants have received their chilling requirement, the plants begin to lose dormancy and become more susceptible to low-temperature injury. For this reason, wide temperature fluctuations in late winter and early spring may result in some degree of winter injury. Low-temperature injury may also result from above-normal fall temperatures that extend into late fall and early winter. Temperatures such as this delay the onset of dormancy, thus making plants more susceptible to low-temperature injury.

The most obvious symptoms of severe low-temperature injury range from death of the entire plant (Plate 133) to death of individual canes or buds (Plate 134) to dieback of individual canes or laterals. Less obvious symptoms occur when vascular tissues are damaged (Plate 135), resulting in a partial impedance of water uptake. In such cases, canes may leaf out and appear normal; however, depending upon the extent of vascular damage, leaves may completely collapse as soon as the plant is stressed by high temperatures or water deficit. If vascular damage is not sufficient to kill plants, they may express less severe symptoms, such as reduced cane growth; small and chlorotic leaves; dry, crumbly fruit; and reduced yield.

Because certain diseases, such as Verticillium wilt and Phytophthora root rot, have aboveground symptoms very similar to some forms of winter injury, it is important that growers can distinguish them. Generally winter injury is not severe enough to kill the roots of affected plants. Even when the injury is sufficient to kill floricanes, the emergence of healthy primocanes from the uninjured roots is common (Plate 133). The emergence of healthy primocanes generally does not occur if the plants have been killed or injured by root-attacking pathogens. In addition, winter injury will generally occur suddenly and is fairly uniform across the planting, whereas damage from root-attacking fungi or other pests will usually occur in localized areas or pockets that tend to spread with time. In several areas of North America, red raspberry plants infected with Phytophthora root rot have mistakenly been diagnosed as having winter injury.

High-Temperature Injury

Excessive heat can also have detrimental effects on *Rubus* plants. The rate of fruit ripening and, ultimately, fruit breakdown or decomposition is increased as the temperature increases. High temperatures and direct exposure to the sun can result in off-colored (white to yellowish), sunscalded fruit. Wilting and eventual collapse and death of plant tissues or the entire plant may occur when the amount of water transpired, or lost by the leaves, exceeds the amount that is taken up by the roots. This can be caused by excessive heat, wind (desiccation), or insufficient soil moisture (drought). Under conditions in which water loss exceeds uptake, the plant "shuts down," closing its stomates and reducing or stopping photosynthesis. This can result in elevated leaf temperatures and, subsequently, reduced plant growth, fruit size, and quality. In general, blackberries tolerate high temperatures better than raspberries.

High soil temperatures can also have negative effects on plant growth. There is no data available specific to *Rubus* spp., but root growth of other fruit crops, such as strawberry and apple, is inhibited at soil temperatures above 30°C.

In locations with warmer climates, such as Australia, New Zealand, and Israel, the prolonged exposure of canes to high temperatures, accompanied by the lack of cold temperatures (chilling), can lead to failure of lateral bud development, because of the persistence of dormancy. This condition is referred to as blind bud. Thus, the selection of species or cultivars with heat tolerance should be accompanied by a low requirement for winter chilling.

Wind

Rubus spp., particularly red and black raspberries, are extremely susceptible to wind injury. Wind can result in three general types of injury: 1) cane breakage, or snapping off of the cane at the base; 2) abrasions on the cane, caused by the cane rubbing against other canes or existing support systems; and 3) desiccation.

Cane breakage is the most obvious form of wind injury. When canes are broken off at the base, they generally wilt and collapse within a few hours. This rapid collapse may appear as a scattered or random event across the planting, or it can be most prevalent on the windward side. When individual canes collapse, they may easily be mistaken for other types of root problems, such as Verticillium wilt, Phytophthora root rot, or damage from cane-girdling insects. However, close examination of the base of the cane reveals that it is broken and is usually being held up by a few fibrous vascular tissues at the base or by surrounding canes. A diagnostic technique to separate wind damage from other forms of cane and root damage is to firmly grasp the collapsed cane (preferably with gloves) and give it a sharp upward pull. If the cane snaps off easily, the problem is likely to be wind breakage at the base. If canes are intact at the base, most people will not be able to easily pull them off, even if they are dead. The use of trellis support systems can increase plant canopy per unit area while simultaneously supporting the canes and reducing wind breakage.

Another common source of wind injury, but not as obvious as cane breakage, is desiccation. High winds, especially during winter, can cause excessive desiccation, resulting in increased low-temperature injury and related cane damage and mortality. One study in England showed that red raspberry plants that were protected by a windscreen (which reduced wind speed from 1.7 to 1.2 m/sec) resulted in plants that were 30% taller and had 40% higher yields. Wind speed directly affects moisture levels in the plant and soil. High winds cause increased evaporation from exposed plant tissues as well as from the soil.

High wind itself will not result in lower ambient air temperature. If it is 0°C, it will be 0°C if the wind is blowing 5 miles an hour or 50 miles an hour. However, winds can affect how rapidly the temperature drops within the bud or cane, and usually it is the rapid or drastic fluctuations in temperature that are the most harmful to the plant.

Windbreaks can provide some degree of protection from wind injury on exposed sites. A good windbreak is a hedge or fence with about 50% permeability. If the site is sloped, the windbreak should run with and not across the slope, to prevent the impedance of air drainage. Generally, a windbreak will provide protection on the leeward side for a distance of up to eight times the height of the windbreak.

Some air movement during the growing season is desirable, however, to reduce the drying time of foliage and fruits. A faster drying time reduces the amount of time that susceptible plant tissues are wet, thus preventing the conditions required for infection by many fungal pathogens.

Soil Moisture

Excessive Soil Moisture

Rubus spp. perform poorly on excessively wet soils; therefore, a first consideration in establishing a commercial *Rubus* planting is to select a site that has good drainage. Poor drainage should be corrected by the use of drain tile or other methods prior to establishing the crop. Planting raspberries or blackberries on poorly drained soils is clearly a poor management decision that generally cannot be corrected after the planting is established.

The reasons for the poor performance of *Rubus* spp. on wet soils are not fully understood; however, wet sites (saturated soil conditions) are conducive to the development of Phytophthora root rot. In addition, excessively wet soils produce anaerobic (low-oxygen) conditions that are not conducive to normal plant growth and health. If soils stay saturated with water for extended periods of time, plants may literally drown. In general, black raspberries are most adversely affected by wet soils; purple and red raspberries and trailing blackberries are somewhat intermediate; and erect blackberries tend to be the most tolerant. Red raspberries appear to be more susceptible to Phytophthora root rot than black raspberries or blackberries.

In general, *Rubus* spp. require deep, well-drained soil with 2–4% organic matter. While they can grow in coarser soil, *Rubus* plants will generally require less supplemental water and nutrients if grown on loamier soils.

One method for assessing soil drainage in prospective sites is to dig several holes, 0.3–1 m deep, around the planting. If the subsoil is black, brown, or reddish brown, the soil is probably well drained. If the soil is yellow or gray, the site probably has poor internal drainage and should be avoided. It may also be useful to fill these holes with water and observe how long it takes them to drain. For good drainage, water should drain out of the hole within 1 hr.

Insufficient Soil Moisture

Cane growth and fruit size of *Rubus* spp. are adversely affected by insufficient soil moisture. The most obvious symptoms of insufficient water are wilting, reduced cane growth and fruit size, and poor-quality (dry and crumbly) fruit. Under severe water deficit, wilting is followed by foliar necrosis and, eventually, death of the plant. The availability of supplemental irrigation is desirable in commercial *Rubus* plantings. In dry seasons, irrigation at the start of fruit ripening can have a dramatic effect on increasing or maintaining acceptable fruit size. Plants that go into dormancy under drought stress are probably much more susceptible to low-temperature injury.

Trickle irrigation can provide the maximum benefits of supplemental water with minimal water usage. In addition, it keeps fruit and foliage dry, thus potentially reducing the incidence of certain fruit-rotting and foliar diseases. Mulches may also prevent damage caused by soil water deficits by moderating rapid and extreme fluctuations in soil water level and temperature. The economics of applying mulch to *Rubus* plantings need careful consideration.

Light

Light is the primary source of energy for the plant. Generally, canes that have better light exposure will bear more and higher-quality fruit. Insufficient light can result in plants that are chlorotic and etiolated. It is important that producers understand the basic need for light in the production of high-quality fruit. Practices that open up the canopy, such as removing old floricanes, thinning excessive canes, and trellising, allow for greater penetration of sunlight and increased productivity and promote faster drying of fruit and foliage. The establishment of plants in areas that are shaded at any time during the day should be avoided.

Selected References

Schoeneweiss, D. F. 1975. Predisposition, stress, and plant disease. Annu. Rev. Phytopathol. 13:193-213.

Vasilakakis, M. D., Struckmeyer, B. E., and Dana, M. N. 1979. Temperature and development of red raspberry flower buds. J. Am. Soc. Hortic. Sci. 104:61-62.

Westwood, M. 1978. Temperate Zone Pomology. W. H. Freeman, San Francisco.

Williams, I. H. 1959. Effects of environment on *Rubus idaeus* L. III. Growth and dormancy of young shoots. J. Hortic. Sci. 34:210-218.

Williams, I. H. 1959. Effects of environment on *Rubus idaeus* L. IV. Flower initiation and development of the inflorescence. J. Hortic. Sci. 34(4):219-228.

(Prepared by B. L. Goulart and M. A. Ellis)

Solar Injury

Sunburn (Sunscald)

Sunburn, or sunscald, results in a section of *Rubus* fruit turning brown and dry (Plate 136). This disorder appears to be more common in blackberries than in raspberries, and erect or semierect blackberries are more prone to sunburn than trailing types. Differences in susceptibility between cultivars within the same species also exist. Berries exposed to direct and intense afternoon sunlight appear to be most susceptible. The exact nature of the disorder is not understood. High temperatures in the absence of direct sunlight also appear to be involved, because berries that are shaded by the canopy can also develop the disorder in hot climates.

White Drupelet Disorder (Solar Injury)

This disorder is observed most commonly on red raspberry, and in certain cultivars during hot growing seasons it can be quite prevalent. The disorder is characterized by drupelets that enlarge normally but fail to turn red (Plate 137). These white drupelets may occur in groups or be randomly distributed over the surface of the fruit. Except for the absence of color, there appears to be no rot or other undesirable characteristic associated with affected drupelets; therefore, the berries are usable for processing. Unfortunately, such berries are unacceptable for the fresh market. The disorder most often occurs in areas in which berries are exposed to high temperatures and intense solar radiation, and it has been identified as a form of solar injury. A. R. Renquist et al recently reported that susceptibility to injury was found to rapidly increase as fruit matured from the green to the white and pink stages. More than two unpigmented drupelets per berry occurred at 42°C and higher with four or more hours of ultraviolet (UV) radiation. Their results also suggested that reducing the level of UV radiation alone, without lowering temperature, is likely to protect raspberries in the field. Thus, shading raspberry plants to reduce UV radiation may lessen the problem. Cultivars apparently differ in susceptibility to the disorder.

Selected References

Renquist, A. R., and Hughes, H. G. 1987. Fruit yield and solar injury comparison for two primocane fruiting red raspberries. Fruit Var. J. 41:133-136.

Renquist, A. R., Hughes, H. G., and Rogoyski, M. K. 1989. Combined high temperature and ultraviolet radiation injury of red raspberry fruit. HortScience 24:597-599.

(Prepared by B. C. Strik)

Part IV. Effects of Cultural Practices on Disease

The use of fungicides for control of several *Rubus* diseases plays an important part in the overall disease management program for *Rubus* fruit crops, but various cultural practices are equally important in achieving effective disease control. Many important diseases, particularly those caused by viruses, cannot be directly controlled by pesticides, and thus their control is almost completely dependent upon the use of cultural practices. An effective disease control program should emphasize the integrated use of specific cultural practices, disease-resistant cultivars, and timely application of pesticides when needed.

Using Disease- and Insect-Resistant Cultivars

The most effective way to avoid pest damage is to use a strongly resistant or tolerant cultivar. Strong resistance is currently available for only a few of the most destructive pests of *Rubus*, but the potential for the introduction of new cultivars with reliable disease and insect resistance is promising. It is important, where possible, to incorporate the use of currently available resistance or tolerance into disease management programs. The most susceptible cultivars should be avoided, especially in situations or locations where a particular disease or insect pest presents a serious risk.

Perhaps the most successful examples of disease resistance achieved in *Rubus* breeding are those effective against viruses and their insect vectors. Two examples illustrate the substantial progress in this area.

In 1937, C. D. Schwartze and G. A. Huber reported that certain red raspberries and their progenies did not support colonization by the large raspberry aphid (*Amphorophora agathonica* Hottes), the principal vector of members of the raspberry mosaic virus complex in North America. This work was later expanded in the United Kingdom by R. L. Knight and his colleagues, who worked with *A. idaei* (previously known as *A. rubi*) and subsequently released numerous cultivars that escaped raspberry mosaic because of resistance to the local aphid vector. This is now the standard approach in several *Rubus* breeding programs. For example, the laboratory of H. A. Daubeny at the Agriculture Canada Research Station in Vancouver is currently the center in North America for breeding for resistance to *A. agathonica*.

Control of the pollen-borne and seedborne raspberry bushy dwarf virus (RBDV) has been approached by the development and use of immune cultivars. By chance Willamette, a major red raspberry cultivar in the Pacific Northwest, is immune to this virus. The inheritance of immunity in raspberry to the common strain of RBDV was studied by D. L. Jennings and colleagues in Scotland. A resistance-breaking (RB) strain of RBDV capable of infecting all but three of the tested cultivars immune to the common strain was subsequently found in the United Kingdom. The RB strain of RBDV has not been reported in North America. Nevertheless, plans are being made to incorporate genes for immunity from the RB strain into new red raspberry cultivars and to attempt, by genetic engineering, to develop cultivars that are protected from RBDV infection by elements of the RBDV genome incorporated into them.

It is beyond the scope of this compendium to thoroughly review the area of disease and insect resistance in *Rubus*, but the subject has been extensively reviewed elsewhere.

Selected References

Jennings, D. L. 1988. Raspberries and blackberries: Their breeding, diseases and growth. Academic Press, New York. 230 pp.

Jennings, D. L., Daubeny, H. A., and Moore, J. N. 1991. Blackberries and raspberries (*Rubus*). Pages 329-390 in: Genetic Resources of Temperate Fruit and Nut Crops. J. N. Moore and J. N. Ballington, eds. International Society for Horticultural Science, The Hague.

Keep, E. 1989. Breeding red raspberry for resistance to diseases and pests. Plant Breed. Rev. 6:245-321.

Excluding and Reducing Pathogen Populations

Growers can employ several cultural practices that can prevent introduction of pathogens into the planting or reduce inoculum levels of pathogens that are endemic to the planting.

Selection and Use of Disease-Free Planting Material

Plantings should be established only from plants certified (indexed) to be free of viruses and other damaging diseases and pests. The importance of selecting healthy planting stock to avoid disease cannot be overemphasized.

Site Selection

Improper site selection can be one of the costliest mistakes a grower can make. Soils infested with long-lived pathogens, such as *Verticillium* or *Phytophthora*, or damaging levels of nematodes should be avoided. Similarly, the establishment of a planting next to an abandoned plantation or an area extensively colonized by wild *Rubus* should be avoided if possible, because these areas serve as reservoirs for many diseases and insect pests.

Site Preparation

Particularly if *Rubus* has been previously planted in the field, crop rotation or fumigation should be considered in order to minimize the carryover of disease inoculum. The eradication of wild *Rubus* in and around the planting site is most effectively performed prior to planting.

Removal of Infected Plants and Tissues

Most fungal diseases overwinter on infected canes within the planting. Old dead floricanes should be removed from the planting as soon as possible during autumn pruning, and heavily infected primocanes should also be removed. All prunings should be removed from the planting and destroyed. Frequent inspections for diseases during the growing season are essential and particularly important for identifying systemic and debilitating diseases, such as viruses, orange rust, and rosette. Once diseased material is discovered, it should be removed immediately from the planting; failure to do so leaves disease reservoirs in the planting.

Eradication of Wild *Rubus* Hosts from Within and Near the Planting

The initial advantages of using expensive disease-free certified planting material are greatly reduced if wild *Rubus* spp. are permitted to grow in adjacent hedgerows, woods, and fields or within the planting. Wild *Rubus* provides an excellent reservoir for viruses and several other important diseases and insects, which are constantly being introduced by birds and other fauna into and around the planting. Therefore, frequent inspection for and removal of wild *Rubus* is highly desirable.

Alternate-Year Bearing (Biennial Cropping)

Most fungal pathogens and some insect pests are specific to *Rubus* spp. With biennial cropping, the complete removal of all overwintering canes and old cane stubs in large blocks of a planting removes all fungal inoculum surviving on canes that is available for infecting next year's primocanes in the spring. In addition, when all floricanes are removed, and all the primocanes produced during the season are retained for cropping in the following season, yields are substantially higher than in annual bearing—provided successive flushes of young canes are removed by desiccant herbicides or by pruning to reduce their competition with floricanes. These higher yields help to partially make up for the loss of a crop.

Biennial cropping is an excellent method for disrupting the disease cycle of several important pathogens by eliminating overwintering inoculum. This system has been shown to reduce the severity of raspberry yellow rust, blackberry purple blotch, and rosette. *Rubus*-specific insect pests, which can only survive by laying eggs on primocanes, should also be controlled if the alternate-year system is started in the fruiting year when all primocanes are removed. Further research over several seasons is required to study the impact of alternate-year bearing on diseases and pests, but it holds the promise of reduced pesticide usage, even if the yield over two years is less than that for annual bearing.

Modifying Microclimate Within the Planting

Microclimate refers to the climate within the canopy of the row or individual plant. In relation to disease, the important elements of canopy microclimate are relative humidity, ventilation or air circulation, the temperature of the air and plant tissues, and the intensity and quality of light. Factors that open the canopy and increase air circulation generally reduce disease incidence and severity by lowering the humidity, shortening periods of wetness, and improving spray penetration and coverage. By employing several cultural practices that can directly or indirectly affect the microclimate within the planting, growers can contribute greatly to the disease management program.

Site Selection

Most commercial *Rubus* cane fruits grow poorly on waterlogged sites. Low-lying, poorly drained sites should be avoided. These sites create anoxic conditions for root growth and are conducive to the development and spread of Phytophthora root rot. Low-lying planting sites that are prone to frost may increase the likelihood of winter injury and increase susceptibility to cane-infecting fungi or other diseases, such as crown gall. Sloping sites encourage free movement of air and enhance ventilation, whereas low-lying sites tend to collect still, moist air and prolong wet periods. Although natural windbreaks can reduce desiccation of foliage, prevent cane breakage, and reduce the severity of winter injury, they can also decrease air circulation in the planting and canopy and thus prolong wetting periods.

Weed Control

Controlling weeds between and, especially, within the row is essential for successful disease control. In addition to competing with plants for light, water, and nutrients, weeds are very effective in reducing air circulation within the planting and thus prolong periods of leaf wetness and infection periods for many important fungal and bacterial diseases. Weed species may serve as important reservoirs of inoculum for viruses, nematodes, and some insects.

Manipulation of the Plant Canopy

Any practice that alters the density of the plant canopy in order to increase air circulation and exposure to sunlight is generally beneficial to disease control. Optimizing between-row and within-row spacings and maintaining interplant spacings through judicious cane thinning throughout the life of the planting are highly desirable. Control of plant vigor, particularly through avoidance of high levels of nitrogenous fertilizers and careful use of cane vigor control techniques, can greatly aid in improving the canopy. Specialized trellis designs for various *Rubus* spp. can further improve air circulation and increased exposure to sunlight and increase harvest efficiency. Trickle irrigation (as opposed to overhead sprinkler irrigation) greatly reduces the wetting of foliage and fruit and the risk of splash dispersal of several important fungal pathogens.

Modifying the Plant Canopy Through Cane Suppression

Studies in the United Kingdom and Poland have shown that the replacement canes produced after the first flush of young primocanes is removed are less severely affected by several fungal diseases and an important insect pest (raspberry cane midge). It is important to note that cane suppression should never be used, except when excessively vigorous and numerous primocanes are produced, and the cultivar is known to tolerate this method of cane management; otherwise, serious damage may result. The first flush of young canes can be removed, either by a desiccant herbicide or by pruning, when they attain a height of approximately 30 cm. Cane removal induces the buds at the crown and on the roots to produce a second flush of young (replacement) canes which, by the time of harvest, are shorter than first-flush canes but develop sufficiently by the end of the growing season to produce the next year's crop.

Studies have shown that cane blight, midge blight, spur blight, cane Botrytis, anthracnose, and raspberry yellow rust are partially controlled by cane suppression. The disease control mechanisms resulting from cane suppression are complex and likely to vary in effectiveness in different localities and under different cultural conditions. Raspberry yellow rust is controlled because of the almost complete destruction of fungal inoculum at the pycnial stage on the first-flush canes; cane blight, by reducing the number of wounds through which the pathogen enters the cane; midge blight, by changing the timing and number of natural splits available for oviposition by the pest; and spur blight and cane Botrytis, by delaying the onset of cane susceptiblity past the period of inoculum availability. As described above, cane suppression also increases ventilation of the canopy and therefore reduces the duration of wetness. In addition, the more open canopy allows for greater sunlight and spray penetration.

Altering Production Practices to Prevent Plant Injury and Infection

Any cultural practice that reduces physical damage to plant tissues will generally be beneficial to disease control. The development of improved mechanical harvesters that do not wound primocanes can greatly reduce the incidence of cane blight and midge blight. Adequate support of thorny primocanes on appropriate trellis structures can prevent the abrasion of canes, foliage, and fruit during windy conditions. The removal of lateral shoots from the lower portions of floricanes may be beneficial in removing berries that might become contaminated with infested soil and thus cause rot problems in storage.

Careful Handling of Fruit During Harvest, Transport, and Storage

For growers, the successful harvest, shipment, and storage of fruit is probably the most critical portion of the production cycle. Improper handling during this critical period can lead to serious losses in quality or loss of the entire harvested crop. A combination of practices that prevent injury to fruit and create a storage environment unfavorable for disease development is essential to disease management and successful production. For a description of these practices, see Botrytis Fruit Rot (Gray Mold) and Blossom Blight.

Selected References

Lawson, H. M., and Wiseman, J. S. 1983. Techniques for the control of cane vigour in red raspberry in Scotland: Effects of timing and frequency of cane removal treatments on growth and yield in cv. Glen Clova. J. Hortic. Sci. 58:247-260.

Waister, P. D., Cormack, M. R., and Sheets, W. A. 1977. Competition between fruiting and vegetative phases of the red raspberry. J. Hortic. Sci. 52:75-85.

Williamson, B., Lawson, H. M., Woodford, J. A. T., Hargreaves, A. J., Wiseman, J. S., and Gordon, S. C. 1979. Vigour control, an integrated approach to cane, pest and disease management in red raspberry (*Rubus idaeus*). Ann. Appl. Biol. 92:359-368.

(Prepared by M. A. Ellis and B. Williamson)

Part V. Development of Healthy Planting Materials

Raspberry Certification Programs in North America

It is imperative to use high-quality, disease-free planting stock when establishing a perennial crop, such as raspberries and blackberries. For this reason, various programs have evolved over the years with the objective of producing pathogen-free planting stock. Such planting stock is usually designated as certified, and the program that produces it is called a certification program. Certification is the final, and usually written, confirmation that the planting material has been produced according to a specific set of rules. Certification programs provide a quality guarantee to the buyer.

Early raspberry certification programs in North America were primarily designed to control raspberry mosaic disease. This disease, transmitted by the large raspberry aphid, was a problem wherever raspberries were grown. Older cultivars, such as Newburgh, Latham, Cuthbert, and Taylor, supported high populations of aphids. Even when new plantings were established with mosaic-free stock, many plants became infected within a year or two, unless extreme care was taken to exclude the virus from the planting stock and new plantings were located away from established plantings to minimize aphid spread. Today, the mosaic disease is rare in some red raspberry-growing areas of North America, such as the Pacific Northwest, because aphid-susceptible cultivars have been replaced with cultivars that are resistant to the aphid vector. Such cultivars as Canby, Haida, Skeena, Nootka, Chilliwack, Comox, and Tulameen are aphid-resistant, and plantings of these cultivars have remained free from mosaic after several years of field exposure.

Even though the raspberry mosaic problem has largely been solved in some areas, the major raspberry-growing provinces (British Columbia and Ontario) and states (New York, Washington, Oregon, and Michigan) have recognized the need for pathogen-free planting stock, and certification programs have been developed to fill this need. These programs differ somewhat in detail, but they all have the common objective of providing planting material free of pathogens (nematodes, fungi, viruses, and bacteria), genetically uniform and true to name, and of good horticultural quality. The specifics of most programs include registration, field inspection, and certification that the material meets certain pest- and disease-tolerance levels. All programs are based on annual field inspections combined with intensive plant and soil sampling and continual laboratory and greenhouse testing (indexing) for viruses. Nuclear stock (indexed primary source plants, frequently from public research agencies) is usually grown in the field for one year, and having met the required tolerances, is recognized as foundation stock. Foundation stock is also grown in the field for only one year and is used exclusively to produce certified stock. Certified stock may be grown at one site for as many years as required, provided it passes an annual inspection. Once certified stock is sold, it is recognized as commercial stock and no longer remains in the program.

Technically, raspberry certification programs are usually organized and administered by the public sector. This is based on the assumption that only the public sector can provide the detached long-range and independent control that is required. Further, the public sector is needed to provide legal authority and a basis for guaranteeing phytosanitary documentation.

In order to function well, certification programs should involve the cooperative effort of plant breeders, horticulturalists, entomologists, and plant pathologists. The closer the cooperation between these professions, the greater the chance of developing a successful certification program.

Selected References

Anonymous. 1987. British Columbia Raspberry Certification Program and Guidelines. British Columbia Ministry of Agriculture and Fisheries, Victoria.

Buonassisi, A. J., Daubeny, H. A., and Peters, B. 1989. The B.C. raspberry certification program. Acta Hortic. 262:175-179.

(Prepared by R. Stace-Smith)

The Raspberry Certification Program in the United Kingdom

The development of a program to provide growers with raspberry planting material substantially free from virus infection was begun in Scotland in the late 1940s. It was the first such program developed for *Rubus*. As the knowledge of raspberry viruses, transmission, and detection has increased, the program, administered by the Scottish Office of the Agriculture and Fisheries Department, has undergone refinements and improvements. It now also includes loganberry and hybrid berries. A similar program operates in England and Wales through the Ministry of Agriculture, Fisheries, and Food.

The two United Kingdom programs are based on the production and maintenance of virus-tested mother stock plants at the Scottish Crop Research Institute (SCRI). SCRI is the sole designated source of virus-tested *Rubus* in the United Kingdom. All plants entering the programs are tested for all known *Rubus* viruses at SCRI. Such tests involve visual assessments for virus symptoms; bioassays, which use herbaceous virus indicator test plants; graft inoculation to the *Rubus* virus indicator *R. occidentalis*; and enzyme-linked immunosorbent assay (ELISA) for raspberry bushy dwarf virus (RBDV). All known *Rubus* viruses should be detected with these four tests. All plants that appear virus-free from these indexing procedures are termed *virus-tested* and are henceforth regarded as mother stock plants. Currently, SCRI maintains more than 100 virus-tested mother stock cultivars and clones of *Rubus*, mostly of red raspberry. This is the largest such collection in Europe and, possibly, the world. Mother stock plants are maintained in soilless compost in an aphid-proof glasshouse and treated with insecticide at regular intervals during the growing season as an extra precaution. Each year, the plants are virus-indexed by ELISA for RBDV and by graft-inoculation to *R. occidentalis* to monitor their continued freedom from viruses thought most likely to reinfect plants. Inspectors from the Scottish Office of the Agriculture and Fisheries Department also monitor plants for freedom from other diseases and pests and for varietal purity. Mother stock plants are not allowed to flower in the glasshouse in order to minimize the risk of infection by the pollen-borne RBDV and to prevent clonal stocks from being contaminated by hybrid seedlings. However, to ensure that the fruiting characteristics of these vegetatively grown plants are not impaired by the occurrence of mutations, each year canes of mother stock plants are allowed to flower in winter in a separate, heated glasshouse and are hand-pollinated. The resultant fruit is inspected for aberrations, and mother plants that produce abnormal fruit are eliminated. In this way, clones of cultivars that seem prone to develop crumbly fruit, a genetic mutation, have been eliminated before being propagated for release to growers.

Roots harvested from individual mother stock plants during winter are used to produce individual foundation stock plants from root cuttings. This first-stage propagation is done by the Scottish College of Agriculture, which produces about 12,000 such plants each year. These foundation stock plants are raised in aphid-proof greenhouses and, in the winter of the year in which they are propagated, are planted in the field at selected farm sites to form raspberry propagation beds. The selected sites must conform to strict regulations, which include 1) testing the soil prior to planting to ensure freedom from the potato cyst nematode and to assess the occurrence of *Longidorus elongatus* and *Pratylenchus penetrans*, and 2) isolation from cultivated and wild *Rubus* and *Rubus* propagation beds of a lower certified grade. With the exception of autumn-fruiting cultivars, plants are not allowed to flower in these propagation beds. During the growing season, inspectors assess the beds on at least two occasions for freedom from pests, diseases (including virus diseases), and weeds and for varietal purity. Crops are graded and awarded certificates based on tolerance limits for pests and diseases. In Scotland, the highest grade certificate (Super Elite) requires all plants to be free from visible virus symptoms, Rubus stunt, and raspberry root rot disease. The second highest grade (Elite) requires all plants to be free from Rubus stunt, soilborne viruses, and raspberry root rot and 99.5% of plants to be free from visible symptoms of infection by other viruses; the third highest grade (Standard) has the same requirements as Elite, except the tolerance level for visible virus symptoms is 98%. All grades must be relatively free from other diseases and from pests. Propagation at each grade is restricted to four years, provided the plants have an unbroken history at that grade. Thereafter, the Super Elite and Elite grades can be entered for certification only at a lower grade. Canes harvested from these sites bear the specific certificate grade and are sold to growers. In legislation to be implemented in Scotland in 1991, it will become illegal to offer for sale raspberry canes that have no certificate.

The certification programs have been a major factor in improving the health and productivity of commercial raspberry plantations in the United Kingdom through the control of the spread and effects of viruses and some fungal pathogens.

Selected References

Anonymous. 1990. Inspection and Certification of Raspberry Plants 1990. Appendix B. Scotl. Dep. Agric. Fish. 8 pp.

Jones, A. T. 1986. Advances in the study, detection and control of viruses and virus diseases of *Rubus*, with particular reference to the United Kingdom. Crop Res. 26:127-171.

(Prepared by A. T. Jones)

Nursery Production of Virus-Free Planting Material

Throughout this compendium, the importance of using disease-free (virus-indexed) nursery stock to establish new plantings has been emphasized. Plants should always be obtained from a reputable source, preferably a nursery that sells plants from certified virus-free stock. Certification is an assurance that the plants used for propagating this stock were tested by indexing and found to be free of common viruses. This section briefly describes how virus-free (indexed) mother plants are produced in a modern nursery in the U.S.

Nursery production starts by obtaining mother stock plants, from which all other plants are eventually propagated. The source of mother plants may include the National Germplasm Repository in Corvallis, Oregon; plant breeders at public institutions; or various state certification programs. Depending upon their source, mother plants may or may not be certified to be free of plant viruses. Upon arrival at the nursery, they are tested (indexed) for the presence of plant viruses either by grafting to sensitive test (indicator) plants, which develop characteristic symptoms if the source plant is virus-infected, or, where applicable, through the use of commercially available enzyme-linked immunosorbent assay (ELISA). In ELISA, a few drops of sap from the leaves of mother plants are screened for the presence of viruses.

If virus is detected, heat treatment is often used to produce a virus-free plant from the infected stock plant. Some viruses can be effectively eliminated from whole plants by heat treatment, but others that are more heat-stable are more difficult to eradicate. For removal of heat-sensitive viruses, infected plants are placed in special chambers, which are maintained at 37°C for several weeks. After heat treatment, small shoot tips or apical meristems are excised and grown into new plants. Tissue culture methods are often employed to regenerate these meristems into plants, because the smaller the meristem, the more likely that the resulting plant will be virus-free. Meristems excised from heat-treated plants are usually less than 1 mm in length. After additional testing to ensure the absence of viruses, these "elite" mother plants are grown in screened cages or screened greenhouses. The cages prevent exposure to virus-transmitting insects. The air intakes of the greenhouse are screened to provide additional protection from insect vectors. The elite mother plants are systematically checked for trueness of cultivar and freedom from viruses. Through the development of fast and reliable tests such as ELISA, modern nurseries have the capability of indexing plants for several important plant viruses. Unfortunately, the specific antisera needed for ELISA testing have not been developed for many of the most important *Rubus* viruses in North America, and the time-consuming method of grafting onto indicator plants is the only routine method currently available to test for their presence.

Approximately 7 million *Rubus* plants are propagated annually in the United States and Canada. Several methods of propagation are used commercially, and depending upon the nursery and the *Rubus* spp. being purchased, growers will receive either dormant suckers, tip-layered canes, or tissue-cultured plants for planting. Regardless of the method of propagation, the importance of propagating nursery stock from virus-free (indexed) mother plants cannot be overemphasized.

(Prepared by P. Ahrens)

International Exchange Regulations for *Rubus* Plant Material

The genus *Rubus* is native to almost all continents and many small islands throughout the world. Certain diseases and pests attacking this genus have limited distribution and have not yet been observed throughout the world. Several important pests of *Rubus* have the potential of causing extreme economic damage if they are introduced into new fruit production areas. Governmental plant protection and quarantine regulations are instituted to prevent these pests from being distributed. In the United States, for example, Rubus stunt is not present, and quarantine regulations are designed to exclude stunt-infected plant material. In the United Kingdom, import requirements demand that *Rubus* plants be certified (i.e., have official verification) to be free from black raspberry latent virus, cherry leaf roll virus, Prunus necrotic ringspot virus, raspberry leaf curl virus (American), and tomato and tobacco ringspot viruses before they can be imported. *Rubus* importation regulations are different for each country, and they are constantly changing as new diseases are documented and new laws are enacted. The department or ministry of agriculture for each country should be contacted for specifics of their most current regulatory information.

As with other crops, visual inspection of *Rubus* plant material by quarantine personnel at ports of entry is useful in detecting most life stages of arthropods and visible symptoms or signs of plant pathogens. However, symptomless material that is colonized by plant pathogens or contains arthropod eggs will generally escape notice. Thus, official verification or phytosanitary certification by inspectors from the point of origin (exporting country) may be required, to assure the quarantine officers at the point of entry (importing country) that the material is clean.

To obtain *Rubus* plant material from a foreign country, a requester usually must obtain an import permit from his own country. The permit describes the pests that must be excluded. The quarantine inspector from the exporting country then examines the material and prepares a phytosanitary certification stating that the particular plant material accommodates the import permit conditions. The material can then be shipped, with copies of the permit and the certification, to the requesting country. Upon arrival, the material is usually examined by plant inspectors and forwarded to the requester or the country's quarantine station as required.

U.S. federal regulations for importation of foreign *Rubus* material are listed in the Plant Protection and Quarantine (PPQ) Programs of the Animal and Plant Health Inspection Service (APHIS) of the Department of Agriculture. These regulations are periodically reviewed and revised. A summary of the U.S. *Rubus* restrictions in effect in 1990 are given in Table 2. Scientists requesting germ plasm for study, individuals interested in obtaining foreign germ plasm for trial propagation, or anyone wanting to import *Rubus* plants must comply with these regulations. Contact the U.S. Department of Agriculture, APHIS, PPQ (Federal Office Building, Hyattsville, MD 20782), to apply for a postentry import permit or for further information concerning federal importation regulations.

Table 2. Summary of U.S. Federal Regulations Concerning *Rubus* Importation

Plant Material	Importation Prohibited	Post-Entry Restriction[a]	Other Requirements
Rubus spp.	Plants from Europe if not certified for Rubus stunt	Plants from all countries (except Canada [Ontario] when certified free of Rubus stunt)	Written permit of certification
R. fruticosus, including seeds			Noxious weed permit
R. moluccanus, including seeds			Noxious weed permit

[a] Post-Entry Quarantine Plants are those listed as "restricted" in the USDA/APHIS Plant Protection and Quarantine Programs, Title 7, Agriculture, Chapter III, Part 319.37. These regulations are revised periodically.

Some states, provinces, or other governmental jurisdictions also impose restrictions on the importation of plant material. Contact your state department of agriculture or corresponding organization to find out which regulations apply locally. In some states, nursery certification against the presence of the brown garden snail is a requirement before *Rubus* or other plants can be shipped across state lines.

Several *Rubus* species, including *R. fruticosus* and *R. molaccanus*, are labeled as noxious weeds in the U.S. Importation of plants or seeds of these species into the U.S. from foreign sources or movement of this germ plasm across state lines requires the permit PPQ form 526. This permit can be requested through APHIS at the above address. Other countries, such as Australia and New Zealand, also consider certain *Rubus* species to be noxious weeds and require permits and great care to prevent escape from cultivation.

Quarantine regulations are critical in preventing the movement of disease agents or pests. As global exchange of germ plasm becomes more common, plant quarantine regulations become increasingly important for the protection of economic crops.

Selected References

APHIS-PPQ. 1980. Foreign quarantine notices. Title 7, Agriculture. Chapter III, Animal and Plant Health Inspection Service. Part 319.37:1-26. Fed. Reg. 45(94):31572-31597.

APHIS-PPQ. 1983. Noxious weeds. Title 7, Agriculture. Chapter III. Part 360. Fed. Reg. 48(8):20037-20040.

Parliman, B. J., and Daubeny, H. A. 1988. Considerations for effective exchange of clonally propagated plant germplasm. HortScience 23(1):67-73.

(Prepared by K. Hummer)

Glossary

a.i.—active ingredient
C—Celsius (°C = (°F −32) × 5/9)
cm—centimeter (1 cm = 0.39 inch; 1 inch = 2.54 cm)
Da—dalton (1 Da = 1.663×10^{-24} g)
g—gram (1 g = 0.3527 ounce; 453.6 g = 1 pound)
ha—hectare (1 ha = 10,000 m^2 = 2.471 acres)
hr—hour
kbp—kilobase pair
kg—kilogram (1 kg = 1,000 g = 2.205 pounds)
L—liter (1 L = 1.057 quarts liquid)
m—meter (1 m = 39.37 inches)
mm—millimeter (1 mm = 0.001 m)
μm—micrometer (1 μm = 10^{-6} m)
M_r—relative molecular mass
nm—nanometer (1 nm = 10^{-9} m)

acervulus (pl. acervuli)—saucer-shaped or cushionlike fungal fruiting body bearing conidiophores, conidia, and sometimes setae
aeciospores—unicellular, nonrepeating spores, usually resulting from dikaryotization, which give rise to dikaryotic mycelium
aecium (pl. aecia)—fruiting body (sorus) of a rust that produces aeciospores
alternate-year bearing—system of cane management that allows a planting to fruit every second year
anamorph—the asexual form (also called the imperfect state) in the life cycle of a fungus, in which asexual spores (such as conidia) or no spores are produced
anastomosis (pl. anastomoses)—fusion, as of hyphal strands, and combination of their contents
anther—the pollen-bearing portion of a stamen
antheridium (pl. antheridia)—the male sex organ of some fungi
anthracnose—disease caused by acervuli-forming fungi (order Myrianiales) and characterized by sunken lesions and necrosis
antibody—a protein formed in the blood of warm-blooded animals in response to the injection of an antigen
antigen—any foreign chemical (normally a protein) that induces antibody formation in animals
aphicide—an insecticide that kills aphids
appressorium (pl. appressoria)—swollen, flattened portion of a fungal filament that adheres to the surface of a host plant, thus providing anchorage for invasion by the fungus
ascocarp—sexual fruiting body (ascus-bearing organ) of an ascomycete
ascogenous—developing or originating from an ascus
ascomycete—member of a class of fungi that produce sexual spores (ascospores) endogenously within an ascus
ascospore—sexual spore borne in an ascus
ascus (pl. asci)—saclike cell in which ascospores (typically eight) are produced
asexual—vegetative; without sex organs, sex cells, or sexual spores, as the anamorph of a fungus
autoecious—producing all spore types on the same host, as a rust fungus
axillary bud—bud that develops in the axil of a leaf (also called a lateral bud; see lateral shoot)

bacilliform—shaped like a blunt, thick rod
bark—brown periderm (a protective tissue) or, more generally, all the tissues outside the cambium (including the phloem and periderm)
basidiomycete—member of a class of fungi that form sexual spores (basidiospores) on a basidium
basidiospore—haploid spore of a basidiomycete
basidium (pl. basidia, adj. basidial)—short, club-shaped fungal cell on which basidiospores are produced
basipetal—downward from the apex toward the base of a shoot or branch; developing in the direction of the base, with the apical part affected first
biennial—a plant that requires two growing seasons to complete its life cycle (vegetative growth in the first year, reproductive growth in the second, followed by plant death)
biflagellate—having two flagella
biological control—disease or pest control through counterbalance by microorganisms and other natural components of the environment
biovar—intrasubspecific group of organisms (usually bacteria) differentiated from other such groups within the same species by biochemical or physiological properties (also called *biotype*)
bitunicate—two-walled, as an ascus
blight—any sudden, severe, and extensive spotting, discoloration, wilting, or destruction of leaves, flowers, stems, or entire plants, usually attacking young, growing tissues (in disease names, often coupled with the name of the affected part of the host, e.g., leaf blight, blossom blight, shoot blight)
budbreak—the stage of bud development when green tissue becomes visible

callus—parenchyma tissue that grows over a wound or graft and protects it against drying or other injury
calyx—the part of a flower consisting of sepals
cambium (pl. cambia)—sheath of meristematic cells in stems and roots, which primarily divide tangentially, producing secondary xylem toward the inside and secondary phloem toward the outside
canker—necrotic, localized diseased area
canopy—the mass of leaf-bearing shoots, measured in height, width, or distribution
causal agent—organism or agent that produces a given disease
certified plants—planting stock produced according to requirements of a program having the objective of producing pathogen-free plants
chlamydospore—thick-walled or double-walled asexual resting spore formed by modification of a hyphal segment
chlorosis (adj. chlorotic)—abnormal light green to yellow plant coloration due to incomplete formation or destruction of chlorophyll
cirrus (pl. cirri)—a tendril-like mass of forced-out spores contained in a mucilage
clavate—club-shaped
cleistothecium (pl. cleistothecia)—closed, usually spherical, ascus-containing structure of a powdery mildew fungus
clone—vegetatively (asexually) propagated plant or member of a group of such plants derived from a single original plant
coalesce—to run together
coenocytic—multinucleate (e.g., pertaining to a multinucleate plant body enclosed within a common wall or a fungal filament lacking cross-walls)
conidiogenous—producing conidia
conidioma—enclosed fungal fruiting body producing conidia
conidiophore—specialized fungal hypha on which conidia (conidiospores) are produced
conidium (pl. conidia)—asexual spore formed by abstriction and detachment of part of a hyphal cell at the end of a conidiophore and germinating by a germ tube
cortex (adj. cortical)—region of parenchyma tissue between the epidermis and the phloem in stems and roots
cross-protection—the phenomenon whereby plants infected with one (usually mild) strain of a virus do not develop additional symptoms when inoculated with a second, severe strain of the same virus
crown—point where the trunk and roots join, at or just below the soil surface
cultivar (abbr. cv.)—a cultivated plant variety or cultural selection propagated vegetatively as a clone
cultural control—the use of production practices to control or improve control of plant pests and diseases

cuticle (adj. cuticular)—water-repellent waxy covering (cutin) of epidermal cells of plant parts, such as leaves, stems, and fruits; the outer sheath or membrane of a nematode
cutin—*see* cuticle

dichotomous—branching, often successively, into two more or less equal arms
dieback (v. die back)—progressive death of shoots, leaves, or roots, beginning at the tips
differentiation—the physiological and morphological changes that occur in a cell, tissue, or organ during development from a juvenile state to a mature state
dormancy—nongrowing condition of a plant, caused by internal factors (as in endodormancy) or environmental factors (as in ectodormancy)
drupelets—the small sections of a *Rubus* fruit, each containing a seed

echinulate—having spines or other sharp projections
ectoparasite—parasite living outside its host
ELISA—*see* enzyme-linked immunosorbent assay
endemic—native to or peculiar to a locality or region
endoparasite—parasite living within its host
enzyme—protein that catalyzes a specific biochemical reaction
enzyme-linked immunosorbent assay (ELISA)—a serological test in which the sensitivity of an antibody-antigen reaction is increased by attaching an enzyme to one of the reactants
epidemic—general and serious outbreak of disease (used loosely with plant diseases)
epidemiology—the study of factors influencing the initiation, development, and spread of infectious disease
epidermis (adj. epidermal)—outermost layer of cells on plant parts
eradicant—chemical used to eliminate a pathogen from a host or an environment
eradicate—to destroy or remove a pest or pathogen after it has caused a disease to become established
erumpent—breaking out or erupting through the surface
estivate—to spend the summer or dry periods in a quiescent or inactive state
extrude—to push out; to emit to the outside
exudate—substance that is excreted or discharged; ooze

fasciation—malformation by which branches, shoots, or floral organs become enlarged and flattened or sometimes curved, as if several parts have been fused
filiform—long, needlelike
flaccid—wilted, lacking turgor
flagellum—hairlike or whiplike appendage of a bacterial cell or fungal zoospore, providing locomotion
floricane—the second-year cane which overwintered and will fruit and die in the current year
floricane-fruiting—forming fruit (as with *Rubus* plants) only on second-year canes in late spring or summer; summer-fruiting
fructification—fruiting body
fruit—mature ovary
fruiting body—any of various complex, spore-bearing fungal structures
fumigant—vapor-active chemical used in the gaseous phase to kill or inhibit the growth of microorganisms or other pests
fungicide (adj. fungicidal)—chemical or physical agent that kills or inhibits the growth of fungi
fusiform—narrowing toward the ends

gall—outgrowth or swelling of unorganized plant cells produced as a result of attack by bacteria, fungi, or other organisms
gelatinous—resembling gelatin or jelly
genome—set or group of chromosomes
genotype—genetic constitution of an individual (in contrast to its appearance, or phenotype)
genus (pl. genera)—group of related species
germ plasm—a collection of genetically diverse plants, including wild material, which can be used to improve cultivated plants through breeding
germ tube—initial hyphal strand from a germinating fungal spore
germinate—to begin growth of a seed or spore
girdle—to circle and cut through; to destroy vascular tissue, as in a canker or knife cut that encircles the stem
gram-negative, gram-positive—pertaining to bacteria that release or retain, respectively, the violet dye in Gram's stain
guttule (adj. guttulate)—oil-like drop inside a spore

hardiness—ability of a plant to withstand cold temperatures
haustorium (pl. haustoria)—specialized outgrowth (of a fungus hypha) that penetrates a host plant and extracts nutrients
herbaceous—nonwoody, as certain plants or plant parts
herbicide—chemical that kills or limits the growth of plants
heteroecious—a rust fungus having spore types produced on two different hosts
heterokaryotic—having two or more genetically different nuclei in the same protoplast
heterothallic—pertaining to species of fungi in which the sexes are separated in different mycelia
hibernaculum—a tent or sheath made of plant material in which an insect larva hides or lies dormant
homothallic—pertaining to species of fungi in which both sexes are present in the same mycelium
host—living plant attacked by or harboring a parasite and from which the invader obtains part or all of its nourishment
hyaline—colorless, transparent
hybrid (v. hybridize)—sexually produced offspring of genetically differing parents (if the parents are of different species, the offspring is an interspecific hybrid; in raspberries and blackberries, further vegetative propagation continues it as a clone)
hyperplasia—abnormal increase in the number of cells in a tissue or organ, resulting in the formation of galls or tumors
hypertrophy—abnormal increase in the size of cells in a tissue or organ, resulting in the formation of galls or tumors
hypha (pl. hyphae, adj. hyphal)—tubular filament of a fungus

immunofluorescence—technique involving the visualization of an antibody-antigen reaction in a light microscope by means of conjugated fluorescent probes
immunosorbent electron microscopy—technique involving the visualization of an antibody-antigen reaction in an electron microscope
imperfect state—*see* anamorph
in vitro—in glass, on artificial media, or in an artificial environment
indexing—determination of the presence of a transmissible disease in a plant
indicator plant—plant that reacts to a pathogen (such as a virus) or to an environmental factor with distinct symptoms, which are used to identify the pathogen or determine the effects of the environmental factor
infection—process in which a pathogen enters, invades, or penetrates and establishes a parasitic relationship with a host plant
infection court—site in or on a host plant where infection can occur
infectious—capable of spreading disease from plant to plant
infective—able to attack a host and cause infection (as in disease-causing organisms or viruses); carrying or containing a pathogen and able to transfer it to a host plant, causing infection (as in vectors)
infest—to attack, as a parasite (especially nematodes); to contaminate, as with microorganisms; to be present in numbers
inoculate (n. inoculation)—to place inoculum in an infection court
inoculum—pathogen or pathogen part (e.g., spore, mycelium) that infects plants
integument—the outer or enveloping layer (as a skin, membrane, or husk) of an organism
internode—the portion of a stem between two adjacent nodes
isolate—pure microbial culture, separated from its natural origin

larva (pl. larvae)—the juvenile or immature stage of a nematode or insect
latent—present but not manifested or visible, as a symptomless infection
lateral bud—bud formed in the axil of a leaf
lateral shoot—a shoot (also called a lateral) produced from a lateral or axillary bud of a leaf
leaf scar—scar left on a stem after leaf fall
lesion—wound or delimited diseased area
locule (adj. locular)—cavity, especially in a fungal stroma

macrocyclic—referring to a long-cycled rust fungus producing at least one type of binucleate spore in addition to the teliospore
matrix—the material in which an organism or organ is embedded
mechanical injury—injury of a plant part by abrasion, mutilation, or wounding
meristem (adj. meristematic)—plant tissue characterized by frequent cell division, producing cells that become differentiated into specialized tissues
mesophyll—central, internal, nonvascular tissue of leaves, consisting of the palisade and spongy mesophyll
microsclerotium (pl. microsclerotia)—microscopic, dense aggregate of darkly pigmented, thick-walled hyphal cells
MLO—*see* mycoplasmalike organisms
mosaic—disease symptom characterized by nonuniform foliage coloration, with a more or less distinct intermingling of normal

and light green or yellowish patches, usually caused by a virus; mottle
mottle—disease symptom characterized by light and dark areas in an irregular pattern on leaves or fruit
mucilaginous—viscous, slimy
mummify—to dry and shrivel up
mummy—a dried and shriveled fruit
mutation—heritable genetic change in a cell or plant
mycelium (pl. mycelia)—mass of hyphae constituting the body (thallus) of a fungus
mycoplasma—member of a group of prokaryotic organisms smaller than conventional bacteria and larger than viruses, without rigid cell walls and variable in shape, reproduced by budding or fission
mycoplasmalike organisms (MLOs)—microorganisms found in phloem tissue that resemble mycoplasmas in all respects except that they cannot yet be grown on artificial nutrient media

necrosis (adj. necrotic)—death of tissue, usually accompanied by darkening to black or brown
nematicide—agent (usually a chemical) that kills or inhibits nematodes
nepovirus—member of a group of nematode-transmitted polyhedral viruses
node—enlarged portion of a shoot at which leaves and buds are located
nonseptate—lacking cross-walls in the fungal filament

obligate parasite—organism that can grow or reproduce only on or in living tissue
oogonium (pl. oogonia)—female egg cell of an oomycete fungus
oospore—thick-walled, sexually derived resting spore of an oomycete fungus
ostiole (adj. ostiolate)—pore; opening by which spores are freed from the papilla or neck of a perithecium or pycnidium
ovary—ovule-bearing portion of a pistil
overwinter—to survive over the winter period
oviposition—the laying of an egg by an insect

palisade—layer or layers of columnar cells rich in chloroplasts, beneath the upper epidermis of plant leaves
papilla (pl. papillae, adj. papillate)—small, round or nipplelike projection
parasite—organism that lives with, in, or on another organism (host) to its own advantage and to the disadvantage of the host
parenchyma—living cells, potentially capable of division, in an organ, such as a root, stem, leaf, or fruit
parthenogenesis—development of an unfertilized egg into a new individual
pathogen (adj. pathogenic)—any disease-producing organism
pedicel—stalk of a flower or fruit
perennial—a plant or plant part which lives for more than two years
periderm—*see* bark
perithecium (pl. perithecia)—a flask-shaped or subglobose, thin-walled ascocarp (fungal fruiting body), containing asci and ascospores and having an ostiole (pore) at the apex, through which spores are expelled or otherwise released
petiole—the stalk of a leaf
pH—negative logarithm of the effective hydrogen concentration, a measure of acidity (pH 7 is neutral; values less than pH 7, acidic; values greater than pH 7, alkaline)
pheromone—a chemical substance used in communication within a species of insect, such that one individual releases the material as a signal and another responds to it
phialide (adj. phialidic)—end cell of a conidiophore; conidiophore of fixed length with one or more open ends, through which a basipetal succession of conidia develops
phloem—food-conducting and food-storing tissue of roots, stems, etc.
phyllody—abnormal formation of leaflike flower petals
physiological race—subdivision within a species, the members of which are alike in morphology but differ from other races in virulence, symptom expression, biochemical and physiological properties, or host range
phytotoxic—harmful to plants (usually used to describe chemicals)
phytotoxicity—injury or damage to a plant due to chemical treatment
pistil—female structure of a flower, composed of stigma, style, and ovary
pith—loose, spongy tissue in the center of a stem
plasmid—a self-replicating piece of DNA that is stably inherited in an extrachromosomal state
primary infection—the first infection of a plant, usually in the spring by a pathogen that has overwintered
primary inoculum—inoculum (usually from an overwintered source) that initiates disease in the field, as opposed to inoculum that spreads disease during the season
primocane—vegetative first-year cane of *Rubus* cane fruit

primocane-fruiting—forming fruit on the tops of first-year raspberry canes (primocanes) near the end of the growing season; fall-fruiting
prokaryotic—lacking a nuclear membrane, mitotic apparatus, and mitochondria
propagule—any part of an organism capable of independent growth
protectant—agent, usually a chemical, applied to a plant surface in advance of a pathogen to prevent infection
protoplasm—living contents of a cell
pseudothecium (pl. pseudothecia)—ascocarp similar to a perithecium and having a dispersed rather than an organized hymenium
pustule—blisterlike, small, erumpent spot, spore mass, or sorus
pycnidiospore—spore (conidium) produced in a pycnidium
pycnidium (pl. pycnidia)—asexual, globose, or flask-shaped fungal fruiting body that produces conidia
pycnium (pl. pycnia)—the pycnidium-like haploid fruiting body of a rust fungus

raster—a complex, in certain beetle larvae, of definitely arranged hairs or spines on the ventral surface of the last abdominal segment, in front of the anus
reservoir—an organism in which a parasite that is pathogenic to other species lives and multiplies without causing damage
resistance (adj. resistant)—property of hosts that prevents or impedes disease development
resting spore—temporarily dormant spore, usually thick-walled, capable of surviving adverse environments often for long periods
rhizomorph—fungal mycelium arranged in strands, rootlike in appearance
rhizosphere—microenvironment in soil near, and influenced by, plant roots
ring spot—disease symptom characterized by yellowish or necrotic rings enclosing green tissue, as in some plant diseases caused by viruses
rot—softening, discoloration, and often disintegration of succulent plant tissue as a result of fungal or bacterial infection
rugose—wrinkled

saprophyte—nonpathogenic organism that obtains nourishment from the products of organic breakdown and decay
sclerotium (pl. sclerotia)—hard, usually darkened and rounded mass of dormant hyphae with differentiated rind and medulla and thick, hard cell walls, which permit survival in adverse environments
secondary infection—infection resulting from the spread of infectious material produced after a primary infection or from other secondary infections without an intervening inactive period
secondary inoculum—inoculum produced by an infection initiated earlier in the same growing season
semiochemical—chemical produced by one organism that incites a response in another organism
senesce (n. senescence, adj. senescent)—to decline with maturity or age, often hastened by stress from environment or disease
septum (pl. septa)—cross-wall
serology—the study of antigen-antibody reactions to detect and identify antigenic substances and the organisms that carry them
shoot—the succulent, green, current-season growth from a bud, including the leaves
shoot tip culture—propagation of growing shoot tips, often done in vitro
sign—indication of disease from direct visibility of the pathogen or its parts
sorus (pl. sori)—compact fruiting structure of a rust fungus
sp. (pl. spp.)—species (sp. used after a genus name refers to an undetermined species; spp. after a genus name refers to several species without naming them individually)
spermagonium (pl. spermagonia)—flask-shaped fungal structure producing sporelike bodies that may function as male gametes (spermatia); pycnium of a rust fungus
spermatium (pl. spermatia)—a sex cell; a nonmatile gamete
sporangiophore—sporangium-bearing body of a fungus
sporangium (pl. sporangia)—fungal structure producing asexual spores, usually zoospores
spore—reproductive body of fungi and other lower plants containing one or more cells; a bacterial cell modified to survive an adverse environment
sporophore—hyphal structure that produces fungal spores
sporulate—to produce spores
stamen—male structure of a flower, composed of a pollen-bearing anther and a filament, or stalk
sterigma (pl. sterigmata)—small, usually pointed protuberance or projection
stigma—structure on which pollen grains germinate in a pistil

stoma (pl. stomata, adj. stomatal)—structure composed of two guard cells and the opening between them, in the epidermis of a leaf, stem, or fruit, functioning in gas exchange

stroma (pl. stromata)—compact mass of mycelium that supports fruiting bodies

subepidermal—beneath the epidermis

subglobose—not quite spherical

substrate—the substance on which an organism lives or from which it obtains nutrients; chemical substance that is acted upon, often by an enzyme

succulent—tender, juicy, or watery (used to describe plant parts)

sucker (sucker shoot)—a young, vegetative cane originating from an adventitious bud on the crown or roots

symptom—indication of disease by reaction of the host

symptomless carrier—a plant infected with a pathogen (usually a virus) but having no obvious symptoms

syn.—synonym

systemic—pertaining to a disease in which the pathogen (or a single infection) spreads generally throughout a plant; pertaining to a chemical that spreads internally through a plant

teleomorph—the sexual form (also called the perfect state, or sexual stage) in the life cycle of a fungus, in which sexual spores (ascospores or basidiospores) are formed after nuclear fission

teliospore—thick-walled resting spore that germinates to form a basidium, produced by some fungi, notably rusts and smuts

telium (pl. telia)—sorus that produces teliospores

tip-layering—method of propagation in which the ends of canes are buried in the soil and new plants develop from them

tipping—removing the tops of primocanes for the purpose of stimulating lateral branching

tissue—group of cells, usually of similar structure, that perform the same or related functions

tolerance—capacity of a plant or crop to sustain disease or endure adverse environment without serious damage, injury, or loss of yield

torus—the center of a *Rubus* fruit to which the drupelets attach; the receptacle

transmit—to spread or transfer, as in spreading an infectious pathogen from plant to plant or from one plant generation to another

unitunicate—having one definable wall or cover

urediniospore—repeating vegetative spore of a rust fungus

uredinium (pl. uredinia)—fruiting body (sorus) of a rust fungus, which produces urediniospores

variegation—pattern of two or more colors in a plant part, as in a green and white leaf

vascular—pertaining to conductive tissues (xylem and phloem)

vascular bundle—strand of conductive tissue, usually composed of xylem and phloem (in leaves, small bundles are called veins)

vector—agent that transmits inoculum and is capable of disseminating disease

vegetative—referring to somatic or asexual parts of a plant, which are not involved in sexual reproduction

veinbanding—discoloration or chlorosis occurring in bands along leaf veins (distinguishing them from interveinal tissue), a symptom of virus diseases

veinclearing—process in which the veins of an infected leaf become translucent

vesicle (adj. vesiculate)—subcellular membranous enclosure

viable (n. viability)—able to germinate, as seeds, fungus spores, or sclerotia; capable of growth

viroid—the smallest known infectious agent, consisting of nucleic acid and lacking the usual protein coat of viruses

virulent—pathogenic; capable of causing disease

viruliferous—virus-carrying (usually applied to insects or nematodes)

water-soaked—wet and dark, usually sunken and translucent lesions or plants

wilt—loss of freshness or drooping of plants due to inadequate water supply or excessive transpiration; a vascular disease interfering with water utilization

witches'-broom—disease symptom characterized by an abnormal, massed, brushlike development of many weak shoots arising at or close to the same point

xylem—water-conducting, food-storing, and supporting tissue of roots, stems, etc.

zonate—marked with zones; having concentric rings, like a target

zoospore—fungal spore with flagella, capable of locomotion in water

Selected References

Agrios, G. N. 1988. Plant Pathology. 3rd ed. Academic Press, New York. 803 pp.

Federation of British Plant Pathologists. 1973. A guide to the use of terms in plant pathology. Phytopathology Paper 17. Commonwealth Mycological Institute, Kew, Surrey, England.

Hawksworth, D. L., Sutton, B. C., and Ainsworth, G. C. 1983. Ainsworth & Bisby's Dictionary of the Fungi. 7th ed. Commonwealth Mycological Institute, Kew, Surrey, England.

Raven, P. H., Evert, R. F., and Curtis, H. 1981. Biology of Plants. 3rd ed. Worth, New York. 686 pp.

Stenesh, J. 1989. Dictionary of Biochemical and Molecular Biology. 2nd ed. John Wiley & Sons, New York. 525 pp.

Walkey, D. G. A. 1985. Applied Plant Virology. John Wiley & Sons, New York. 329 pp.

Index

Acalitus essigi, 71
Aesculus, 52
Agrilus ruficollis, 75, 77; Pls. 115, 116
Agrobacterium, 39, 40
   radiobacter, 39, 40
     strain K84, 40
   rhizogenes, 42
   rubi, 39
   tumefaciens, 39, 40
Akala berry, 1
Allygus mayri, 47
Alpine mosaic agent, 56; Pl. 76
Alternaria, 7, 24
   humicola, 24
Alternaria rot, 24
Alternate-year bearing, 11, 87
Aluminum toxicity, 79
American spruce-raspberry rust. See Late leaf rust
Amide, 82
Amphisphaeriaceae, 12
Amphorophora, 43–44
   agathonica, 43, 44, 46, 71, 86; Pl. 107
   idaei, 44, 46, 71, 86
   parviflorii, 58
   rubi, 44, 86
   sensoriata, 71
AMV. See Arabis mosaic virus
Anthonomus signatus, 69; Pl. 102
Anthracnose, 3–5, 19; Pls. 1–4
   control by cane suppression, 87
   and Septoria leaf spot of blackberry, compared, 18–19
Aphid-resistant cultivars, 44
Aphids, diseases transmitted by, 43–46
Aphis
   idaei, 46
   rubicola, 45, 71; Pl. 108
APHIS Plant Protection and Quarantine Programs, 91
Apium, 46
ApMV. See Apple mosaic virus
Apple mosaic virus, 52; Pl. 71
Arabis mosaic/strawberry latent ringspot nepovirus, 60
Arabis mosaic virus, 48–49; Pl. 62
Arctic raspberry. See *Rubus arcticus*
Arctic rusts, 33
Argyrotaenia
   citrana, 72, 78; Pls. 109, 110
     and western winter moth, compared, 74
   velutinana, Pl. 111
Armillaria, 37–38; Pls. 47, 48
   limonea, 38
   luteobubalina, 38
   mellea, 38
   novae-zelandiae, 38
   ostoyae, 38
   tabescens, 38
Armillaria root rot, 37–38
Arthropods, 63
Arthuriomyces peckianus, 26, 27, 30
Ascochyta, 19
Ascospora, 12

   ruborum, 12
Ascospora dieback, 12; Pl. 14
Aulacaspis rosae, 77; Pl. 121
Aulocorthum solani, 44
Autumn rust. See Late leaf rust

Bacillus popilliae, 65
BCV. See Blackberry calico virus
Bean yellow mosaic virus, 56
Benomyl, 5, 6, 7
Betula, 52
   alba, 13
Biennial cropping, 11, 87
Biovar, 39
Bipyridilium, 82; Pl. 129
Black raspberry, 1, 26. See also *Rubus occidentalis*
Black raspberry latent virus, 56, 91
   and tobacco streak virus, *Rubus* strain, compared, 55, 56
Black raspberry necrosis virus, 43, 44, 51
Black raspberry streak, 56
Black vine weevil, 64; Pl. 88
Blackberry, 1, 2, 23, 25, 26, 40, 74. See also *Rubus* subgenera, *Eubatus*; *R. fruticosus*
Blackberry aphid. See *Amphorophora rubi*
Blackberry calico virus, 53; Pl. 72
   and wineberry latent virus, 53, 55
Blackberry leaf miner, 74
Blackberry psyllid, 73–74; Pls. 113, 114
Blackberry-raspberry hybrids, 2
Blackberry rust, 32–33; Pl. 41
   and anthracnose, compared, 3
Blind bud, 84
Blossom blight, 21–23
Blue mold, 24
Blue stripe wilt, 36. See also Verticillium wilt
Bluestem, 36, 56. See also Verticillium wilt
Boron deficiency, 80
Botryosphaeria cane canker of blackberry, 12–13
Botryosphaeria dothidea, 13; Pl. 15
Botryosphaeria fruit rot, 13
Botryotinia fuckeliana, 10, 21
Botrytis, 10
   cinerea, 4, 8, 10, 11, 17, 21, 22, 23, 24; Pls. 10–12, 26, 27
Botrytis fruit rot, 18, 19, 21–23, 25. See also Gray mold
Botrytis gray mold. See Botrytis fruit rot
Boysenberry, 2, 15, 25, 40, 51, 55, 72; Pls. 19, 21
Boysenberry decline, 56; Pls. 78–80
Braconid wasps, 74
Bramble, 1
Bramble yellow mosaic virus, 56
Brown garden snail, 91
BrYMV. See Bramble yellow mosaic virus
Byturus
   tomentosus, 67, 68
   unicolor, 68; Pls. 99, 100

Calcium deficiency, 79, 80
Cane and leaf rust, 28; Pls. 32, 33

Cane and leaf spot, 18
Cane blight, 5–6; Pls. 5, 6
   control of, 87, 88
   and midge blight, compared, 7
Cane Botrytis, 10–11
   control of, 87
   and gray mold, 21
   and spur blight, compared, 8
Cane breakage, 84
Cane gall, 39–40
Cane rust, 28. See also Yellow rust
Cane spot, 3
Cane vigor control (cane burning)
   cane blight, 6
   midge blight, 7
Canker of apples, 13
Captan, 4
Carbendazim, 5
Carlavirus group, 53
Cascadeberry, 25
Catharanthus roseus, 55
Cerambycidae, 76
Cercospora rubi, 18
Cercosporella rubi, 14; Pls. 16–18
Certification programs. See *Rubus* certification programs
Certified planting stock, 49, 50, 86, 89, 90
Chenopodium, 51, 55, 56
   amaranticolor, 48, 54, 55, 56; Pl. 75
   foetidum, 54
   quinoa, 46, 50, 51, 52, 54, 56, 57, 58; Pls. 66, 70
Cherry leaf roll virus, 53–54, 91; Pl. 73
Cherry rasp leaf virus, 56–57
Chinese raspberry, 1
Chondrostereum purpureum, 20
Choristoneura rosaceana, 72
Chrysanthemum, 46
Cladosporium, 22, 24, 25
   cladosporioides, 24
   herbarum, 24
Cladosporium rot, 24
Clay-colored weevil. See *Otiorhynchus singularis*
Clethridium corticola, 12; Pl. 14
Climbing cutworms, 73; Pl. 112
Clipper, 69; Pl. 102
Cloudberry, 1
CLRV. See Cherry leaf roll virus
CMV. See Cucumber mosaic virus
Colletotrichum gloeosporioides, 24
Colletotrichum rot, 24–25
Common green capsid, 66
Coniothyrium fuckelii, 5, 7
Contaminants of mechanically harvested fruit, 64, 78
Copper deficiency, 80
   and phosphorus deficiency, 79
Cornus nuttallii, 54
Corynebacterium fascians, 42
Coryneopsis, 12
Coryneum ruborum, 12
Cotinis nitida, 65–66; Pl. 92
Criconema, 62

Criconemoides, 62
CRLV. See Cherry rasp leaf virus
Crown borer infestations and Botryosphaeria cane canker, compared, 13
Crown gall, 39–40; Pls. 50, 51
 and leafy gall, compared, 42
Cucumber mosaic virus, 85–86; Pl. 59
Cucumis sativus, 50, 51, 52, 54, 56, 58
Cucumovirus group, 46
Cutworms, 78
Cydonia oblonga, 51
Cylindrocarpon ianthothele, 20
 var. *ianthothele*, 20
Cylindrosporium rubi, 18

2,4-D, 83; Pl. 131
Dagger nematode, 48–49, 50, 60–61. See also *Xiphinema*
Desiccation, 84
Dewberries, 13, 14, 25, 26
Diaspididae, 77
Dicamba injury, 83; Pl. 132
Dicarboximide fungicides, 5
Dichlobenil, 81–82
Dichlofluanid, 4
Didymella applanata, 7, 8, 9, 10, 41
Digger wasps, 77
Dinitroaniline, 82
Diphenyl ethers, 82
Diuron, 82
Dothideaceae, 12
Double blossom. See Rosette
Double dart moth, 73
Downy mildew, 15–16, 44; Pls. 19, 21
Dryberry, 15, 70; Pl. 21
Dryberry mite, 70; Pls. 105, 106

Ectoparasites, 61
Elsinoe veneta, 3
Environmental stress, 83–85
Ericaceae, 49
Eriophyid mites, 70
Erwinia amylovora, 40; Pls. 52, 53
European blackberry. See *Rubus fruticosus*
European dewberry, 75
European earwig, 78
European large raspberry aphid, 71
Euscelis plebeja, 47
Evergreen blackberry, 25, 71

Ferbam, 4
Filipendula ulmaria, 75
Fire blight, 40–41; Pls. 52, 53
Floricanes, 2
Forficula auricularia, 78
Fragaria, 46
 vesca, 51, 56
  var. *semperflorens*, 46, 56; Pl. 76
Frankliniella occidentalis, 1
Fusarium, 7, 20
 avenaceum, 7
 culmorum, 7

Glischrochilus
 fasciatus, 67
 quadrisignatus, 67; Pl. 98
Glufosinate injury, 82–83; Pl. 130
Glyphosate injury, 82, 83; Pls. 126–128
Gnomonia cane canker, 19–20
Gnomonia
 depressula, 19
 rubi, 19–20
Gomphrena globosa, 55, 56
Gooseberry, 74
Graphiphora augur, 73
Gray bark phase, in anthracnose, 3
Gray mold, 21–23; Pl. 26. See also Botrytis fruit rot
 fungicides used to control, 4–5
Gray mold wilt. See Cane Botrytis

Green June beetle, 65–66; Pl. 92
Green muscadine fungus, 66
Gymnoconia
 nitens, 26, 27, 28, 30
 peckiana, 26

Hairy root, 42
Halo blight, 41
Hamaspora longissima, 34
Hapalosphaeria deformans, 25
Hemicycliophora, 62
Hendersonia rubi, 12
Herbicide injury, 80–81
Herbicides
 chemical properties of, 81
 contact, 82–83
 postemergent, 82–83
 preemergent, 81–82
 translocatable, 82, 83
 volatile, 83
High-temperature injury, 83, 84
Highbush blackberry, 2
Himalaya berry, 70
Himalaya blackberry. See *Rubus procerus*
Honey fungus, 38
Humulus, 52

Ilarvirus group, 52, 54
Illinoia, 58
 davidsonii, 58
 rubicola, 44
Importation of *Rubus* plant material, restrictions on, 91
Insects, 63
Iprodione, 5
Iron deficiency, 79, 80

Japanese beetle, 65; Pl. 91
Japanese wineberry. See *Rubus phoenicolasius*

Kuehneola uredinis, 28, 33; Pl. 33

Lachnocladiaceae, 38
Lampronia rubiella, 67; Pls. 96, 97
Large raspberry aphid, 43, 71, 78, 86, 89; Pl. 107
Late leaf rust, 30–32, 33; Pl. 37
 and orange rust, compared, 30
Late raspberry rust. See Late leaf rust
Late yellow rust. See Late leaf rust
Leaf miners, 74
Leaf rollers, 72
Leaf-spotting mosaic, 43
Leafhoppers, diseases transmitted by, 46–47
Leafy gall, 42
Leak disease, 23–24
Lepidopteran larvae, 78
Leptosphaeria coniothyrium, 5, 6, 7, 29, 76; Pl. 5
Lesion nematode, 61
Lesser peach tree borer, 63
Light
 and plant nutrition, 79
 and fruit production, 85
Limberneck, 75
Loganberry, 2, 15, 23, 25, 34, 51, 55, 67, 68, 70, 74, 75, 76, 89
Loganberry cane fly, 75
Loganberry degeneration, 51
Long-horned beetles, 76
Longidorus, 48, 49, 60–61
 attenuatus, 49, 60
 diadecturus, 60
 elongatus, 49, 60, 90
 macrosoma, 49, 60
Low-temperature injury, 83–84; Pls. 133–135
 and Botryosphaeria cane canker, compared, 13
 and Phytophthora root rot, compared, 84

Lycocoris pabulinus, 66
Lygus, 78
 lineolaris, 66–67; Pls. 94, 95
Lygus bugs, 66–67

Macrodactylus subspinosus, 66; Pl. 93
Macropsis, 47
 fuscula, 46–47
Macrosiphum
 euphorbiae, 44, 55
 fragariae, 44
Magnesium deficiency, 79–80
Malus, 52
Manganese deficiency, 79, 80
Manganese toxicity, 79
Masonaphis rubicola, 44
MBC-generating fungicides, 5
Meloidogyne, 62
 hapla, 62
Metallus
 pumilus, 74
 rubi, 74
Metarrhizium anisopliae, 66
Midge blight, 7, 75, 76; Pl. 118
 control of, 7, 87
Milky spore disease, 65
Mites, 63. See also Raspberry leaf and bud mite; Raspberry mite; Spider mites
Molybdenum deficiency, 80
Monophadnoides geniculatus, 74
Mother stock plants, 90
Mountain blackberry, 2
Mucor, 23–24
 hiemalis, 23
 piriformis, 23
Mycosphaerella
 confusa, 18
 rubi, 19
Myzus ornatus, 44

Napropamide, 82
Nectria, 20
 mammoidea var. *rubi*, 20
 rubi, 20
Nectria canker of raspberry, 20
Needle nematodes. See *Longidorus*
Nematodes, 59–61
 diseases transmitted by, 47–50
Nepoviruses, 50, 58, 60, 61
 European, 47–49
Nicotiana
 clevelandii, 46, 51, 57
 occidentalis subsp. *obliqua*, 53
 tabacum, 48, 50, 54, 56, 58; Pls. 69, 74
Nitidulidae, 67
Nitrile, 81–82
Nitrogen
 deficiency, 79
 toxicity, 79
Noctuidae, 73, 78
Norflurazon injury, 82; Pl. 125
Nutrient deficiencies, 79, 83
Nutritional disorders, 79–80

Oak fungus, 38
Oak root fungus, 38
Oberea bimaculata, 76–77; Pl. 119
Oblique-banded leaf roller, 72
Obscure root weevil, 64
Oecanthus nigricornis, 77
Oestlundia rubicola, 44
Ollalieberry, 15
Olpidium brassicae, 58
Omphalospora, 12
Operophtera, 74
 bruceata, 74
 brumata, 74
 danbyi, 74
 occidentalis, 74
Orange-banded digger wasp, 66

Orange rust, 26–27, 87; Pls. 30, 31
  and cane and leaf rust, compared, 28
  and late leaf rust, compared, 30
Orange tortrix. *See Argyrotaenia citrana*
Oryzalin, 82
Otiorhynchus, 64
  ovatus, 64
  singularis, 64; Pl. 89
  sulcatus, 64; Pl. 88
Oxyfluorfen, 82

Papaipema nebris, 78
Paraquat injury, 82; Pl. 129
Paravespula, 69
  germanica, 69
  pensylvanica, 69
Peach rosette virus, 60
Peach tree borer, 63
Pegomya rubivora, 75
Penicillium, 24
Penicillium rot, 24
Pennisetia
  bohemica, 63
  hylaeiformis, 63
  marginata, 63; Pl. 86
Peridroma saucia, 73
Peronospora
  rubi, 15
  sparsa, 15–16; Pls. 19, 20
Phaseolus vulgaris, 51, 52, 56
Philaenus spumarius, 47
Phoma, 8
  macrostoma var. *macrostoma*, 7
Phosphorous deficiency, 79
Phosphorous toxicity, 79
Phragmidium, 34
  arcticum, 34
  bulbosum, 33
  imitans, 29
  rubi-idaei, 29, 30, 34; Pl. 34
  violaceum, 3, 33; Pls. 40, 41
Phyllocoptes gracilis, 70
Phytophthora, 34–36, 86
  cactorum, 35
  cambivora, 35
  citricola, 35
  cryptogea, 35
  drechsleri, 35
  erythroseptica, 34
  fragariae, 34–36
    var. *rubi*, 34
  megasperma, 34, 35, 36
    var. *megasperma*, 35
Phytophthora root rot, 34–36; Pls. 42–45
  and cane breakage, compared, 84
  and tomato ringspot, compared, 50
  and wet soils, 84–85
  and winter injury, compared, 84
Phytoseius macrophilis, 70
Picea
  engelmanii, 31
  glauca, 31, 33; Pl. 39
Picnic beetle, 67–68; Pl. 98
Pin mold, 23
Pollen, diseases transmitted by, 51–52
Popillia japonica, 65; Pl. 91
Postharvest fruit rot, 22–23
Postharvest gray mold, 22
Postharvest soft rot, 23–24
Potassium deficiency, 79
Powdery mildew, 16–17; Pls. 22, 23
Pratylenchus, 59
  crenatus, 59
  penetrans, 40, 59, 61, 90; Pls. 83, 84
    and *Xiphinema bakeri*, compared, 60
  pratensis, 59
  vulnus, 59
Primocanes, 2
Priophorus morio, 74
Prunus, 52

Prunus necrotic ringspot virus, 91
Pseudomonas blight, 41; Pl. 54
Pseudomonas syringae, 41
Pucciniastrum
  americanum, 29, 31, 32, 33; Pls. 37–39
  arcticum, 31, 33
Purple blotch, 11; Pl. 13
Purple raspberries, 2
  and orange rust, 27
Pyridazinone, 82

Raspberry. *See also Rubus idaeus*
  general description of, 1, 2
Raspberry aphids, 71–72, 86
Raspberry beetle, 68
Raspberry bud moth, 67; Pls. 96, 97
Raspberry bushy dwarf virus, 51–52, 89–90; Pls. 69, 70
  and calico symptoms, 53, 55
  resistance-breaking strain of, 51, 52, 86
  and tobacco streak virus, 54
Raspberry calico, 53
Raspberry cane borer, 76–77; Pl. 119
  and Pseudomonas blight, compared, 41
Raspberry cane maggot, 75
Raspberry cane midge, 7, 75–76, 87; Pls. 117, 118
  and *Didymella applanata*, 8
Raspberry crown borer, 63; Pls. 86, 87. *See also Pennisetia*
  and strawberry crown moth, 64, 65
Raspberry fruitworm, 67, 68; Pls. 99, 100
Raspberry hybrids, 1
Raspberry leaf and bud mite, 70; Pls. 105, 106
Raspberry leaf curl disease (Scotland), 48
Raspberry leaf curl virus, 45, 91; Pls. 57, 58
  and *Aphis rubicola*, 45, 71
Raspberry leaf-mining sawfly, 74
Raspberry leaf mottle virus, 43, 44
Raspberry leaf spot, 18; Pl. 24
  and Septoria leaf spot of blackberry, compared, 19
Raspberry leaf spot virus, 43–44
Raspberry maggot, 75
Raspberry mite, 70
Raspberry mosaic disease, 89; Pl. 55
Raspberry mosaic disease complex, 43–44
  and raspberry aphids, 71, 86
Raspberry moth, 67
Raspberry ringspot virus, 48–49
  and *Longidorus*, 60
Raspberry sawfly, 74
Raspberry vein chlorosis virus, 46; Pl. 60
  and *Aphis idaei*, 71
Raspberry yellow dwarf, 48
Raspberry yellow rust, 87
Raspberry yellow spot, 57
RBDV. *See* Raspberry bushy dwarf virus
RBDV-RB. *See* Raspberry bushy dwarf virus, resistance-breaking strain of
RCSV. *See* Rubus Chinese seedborne virus
Red-banded leaf roller, Pl. 111
Red raspberry, 25, 34. *See also Rubus idaeus*
  and orange rust, 27
Redberry, 71
Redberry mite, 71
Rednecked cane borer, 75; Pls. 115, 116
  and *Oberea bimaculata*, compared, 77
Regulations, governmental plant protection and quarantine, 90–91
Resseliella theobaldi, 7, 8, 75–76, 87; Pls. 117, 118
Rhizobiaceae, 39
Rhizopus, 22, 23–24
  nigricans, 23
  sexualis, 23–24
  stolonifer, 23–24
RLCV. *See* Raspberry leaf curl virus
RLMV. *See* Raspberry leaf mottle virus

RLSV. *See* Raspberry leaf spot virus
Root-lesion nematode, 59–60. *See also Pratylenchus*
  and dagger and needle nematodes, compared, 61
Root weevils, 64, 78
Rosa, 52
Rosaceae, 1, 49, 52
Rose bug, 664
Rose chafer, 66; Pl. 93
Rose scale, 77–78; Pl. 121
Rosette (double blossom), 13–15, 87; Pls. 16–18
RRV. *See* Raspberry ringspot virus
Rubus, 1
  allegheniensis, 2
  arcticus, 1, 31, 33, 34
  argutus 2
  baileyanus, 2
  caesius, 75
  canadensis, 2
  chamaemorus, 1
  coreanus, 1, 9, 17
  crataegifolius, 9, 24
    and cane Botrytis resistance, 11
  cuneifolius 2
  ellipticus 1, 57
  frondosus, 2
  fruticosus, 3, 34, 70, 91
  glaucus, 1
  hawaiiensis, 1
  henryi, 54, 56, 57
  hispidus, 2
  idaeus, 1, 51
    subsp. *melanolasius*, 31
    subsp. *strigosus*, 1, 9, 29, 31, 36, 51
    subsp. *vulgatus*, 1, 9, 29, 36
  illecebrosus, 1
  kuntzeanus, 1
  laciniatus, 2, 32, 47, 53; Pl. 61
  lasiocarpus, 9, 34
  leucodermis, 1, 31, 51, 57
  macraei, 1
  macropetalus, 2
  molaccanus, 51, 91
  neglectus, 2
  nitidioides, 2
  niveus, 1, 9
  occidentalis, 1, 3, 9, 31, 34, 51, 56, 89; Pl. 77
    and cane Botrytis resistance, 11
  parviflorus, 1, 19, 51, 52; Pl. 81
  phoenicolasius, 1, 44, 45, 51, 54, 55
  pileatus, 6, 9
    and cane Botrytis resistance, 11
  procerus, 2, 46, 48, 50, 53, 54
  pubescens, 33
  rigidus, 56
  rusticanus, 2
  saxatilis, 33
  spectabilis, 1
  stellatus, 1
  strigosus, 57
  theobaldi, 7
  thyrsiger, 2
  triflorus, 33
  trivialis, 2; Pl. 25
  ursinus, 2, 31, 51, 53, 54, 55
Rubus certification programs, 44, 46
  in North America, 89
  in the United Kingdom, 89–90
Rubus Chinese seedborne virus, 57
Rubus stunt, 46–47; Pl. 61
  quarantine of, 91
Rubus subgenera
  Cylactis, 1, 33
  Eubatus, 1, 2, 12
  Idaeobatus, 1, 2
Rubus yellow net virus, 43; Pl. 56
RVCV. *See* Raspberry vein chlorosis virus

Salmonberry, 1
Sand blackberry, 2
Sap beetles, 67
Scarab beetles, 65–66
Sciopithes obscurus, 64
Scolia dubia, 66
Seedborne dsRNA in wild *Rubus*, 57
Seimatosporium, 12
   lichenicola, 12
Septocyta ruborum, 11; Pl. 13
Septoria, 19
   brevispora, 18
   darrowii, 18, 19
   rubi, 18, 19
      var. *brevispora*, 18
Septoria leaf spot, 18–19; Pl. 25
   and raspberry leaf spot, compared, 18
Sesiidae, 11, 64
Shoestring fungus, 38
Silver leaf, 20
Simazine injury, 81; Pl. 122
Sitobion fragariae, 44
SLRV. *See* Strawberry latent ringspot virus
Small raspberry aphid, 45, 71; Pl. 108
Small raspberry sawfly, 74
Soil, nutrient content of, 79
Soil moisture, 79
   insufficient, 85
   excessive, 84–85
Soil pH, 79
Soil type, 79
Solar injury, 85; Pl. 137
Sorbus, 74
   aucuparia, 13
Sphaceloma necator, 3
Sphaerella, 19
Sphaerotheca macularis, 17; Pl. 22
Sphaerulina rubi, 18, 19; Pl. 24
Spider mites, 69–70; Pls. 103, 104
Spotted cutworm, 73
Spur blight, 7–9, 87; Pls. 7–9
   and Botryosphaeria cane canker of blackberry, compared, 13
   and cane Botrytis, compared, 10
Stalk borer, 78
Stamen blight, 25–26
Stellaria media, 49
Stereum purpureum, 20
Stinkbugs, 78
Strawberry bud weevil, 69; Pl. 102
Strawberry crown borer, 63
Strawberry crown moth, 64–65; Pl. 90
Strawberry latent ringspot virus, 48–49
   and Rubus Chinese seedborne virus, compared, 57

Strawberry raspberry, 1
Strawberry root weevil, 64
Striped tree cricket, 77
Subtropical rust, 34
Sulfur deficiency, 79
Sunberry, 75
Sunburn (sunscald), 85; Pl. 136
Sydowiella cane canker, 19–20
Sydowiella depressula, 19–20
Synanthedon bibionipennis, 64 65; Pl. 90

Taraxacum officinale, 49, 56
Tarnished plant bug, 66–67; Pls. 94, 95
Tayberry, 2, 68, 70, 75, 76; Pls. 2, 106
TBRV. *See* Tomato black ring
Temperature, 79
Tenthredinidae, 74
Terbacil injury, 81; Pls. 123, 124
Tetranychid mites, 69
Tetranychidae, 69
Tetranychus urticae, 69
Thimbleberry, 1, 19, 51, 52, 70; Pl. 81.
Thimbleberry ringspot virus, 57–58; Pl. 81
Thiophanate-methyl, 5, 6
Thrips tabaci, 55
ThRSV. *See* Thimbleberry ringspot virus
Tiphia
   popilliavora, 65
   vernalis, 65
Tobacco necrosis virus, 58
Tobacco rattle virus, 58
Tobacco ringspot virus, 58, 91; Pl. 82
Tobacco streak virus, 54–55, 56
   Rubus strain, 54–55; Pl. 74
   and black raspberry latent virus, compared, 55
   and black raspberry streak, compared, 56
TobRSV. *See* Tobacco ringspot virus
Tomato black ring, 48–49, 60
Tomato ringspot virus, 49–50, 60, 91; Pls. 63–69
TomRSV. *See* Tomato ringspot virus
Tortrix moths, 72
Tree crickets, 77; Pl. 120
Triazine, 81
Trichodorus, 58
Trifolium, 46
Trioza tripunctata, 73–74; Pls. 113, 114
TSV-R. *See* Tobacco streak virus, *Rubus* strain
Tummelberry, 15
Tumor-inducing (Ti) plasmid, 39
Two-spotted spider mite, 69
Typhlodromus pyri, 70

Uracil, 81
Urea, 82

Vararia spp, 38; Pl. 49
Variegated cutworm, 73
Veinbanding mosaic, 43
Verticillium, 37
   albo-atrum, 37
   dahliae, 37
Verticillium wilt, 36–37, 56; Pl. 46
   and cane breakage, compared, 84
   and winter injury, compared, 84
Vespa, 69
   vulgaris, 69
Vigna unguiculata, 52
Vinclozolin, 5
Virus-indexed plants, 90

Wasps, 69
Western winter moth, 74
Western yellow rust, 28. *See also* Yellow rust
White drupelet disorder, 85; Pl 137
White root rot, 38–39
Wind injury, 84
Wineberry, 40, 55. *See also Rubus phoenicolasius*
Wineberry latent virus, 53, 55; Pl. 75
Winter injury. *See* Low-temperature injury
Witches'-broom, 14, 25, 47, 59; Pls. 16, 61
WLV. *See* Wineberry latent virus

Xestia, 73
Xiphinema, 48, 50, 54, 60–61
   americanum, 50, 60–61
   bakeri, 60, 61; Pl. 85
   bricolensis, 61
   californicum, 61
   diversicaudatum, 48, 60
   occidium, 61
   pachtaicum, 61
   pacificum, 61
   rivesi, 50, 61
   thornei, 61

Yellow rust, 28–30; Pls. 34–36
Yellowjacket, 69; Pl. 101
   and raspberry crown borer, compared, 63
Youngberry, 2, 15, 25, 26

Zinc deficiency, 80
   and excess phosphorus, 79
   and high soil pH, 79